Lichens

Lichens
Toward a Minimal Resistance

Vincent Zonca

Translated by Jody Gladding

Preface by Emanuele Coccia

polity

Originally published in French as *Lichens. Pour une résistance minimale*.
© Le Pommier/Humensis, 2021

This English edition © Polity Press, 2023

Excerpt from Hans Magnus Enzensberger, *Die Furie des Verschwindens. Gedichte*
© SuhrkampVerlag Frankfurt am Main 1980. All rights reserved by and controlled through Suhrkamp Verlag Berlin.

Excerpt from Hans Magnus Enzensberger, *Blindenschrift* © Suhrkamp Verlag Frankfurt am Main 1964. All rights reserved by and controlled through Suhrkamp Verlag Berlin.

Excerpt from Saint-John Perse, *Vents* © Editions Gallimard 1946.

Polity Press
65 Bridge Street
Cambridge CB2 1UR, UK

Polity Press
111 River Street
Hoboken, NJ 07030, USA

All rights reserved. Except for the quotation of short passages for the purpose of criticism and review, no part of this publication may be reproduced, stored in a retrieval system or transmitted, in any form or by any means, electronic, mechanical, photocopying, recording or otherwise, without the prior permission of the publisher.

ISBN-13: 978-1-5095-5344-0
ISBN-13: 978-1-5095-5345-7 (paperback)

A catalogue record for this book is available from the British Library.

Library of Congress Control Number: 2022936786

Typeset in 10 on 14 Fournier
by Fakenham Prepress Solutions, Fakenham, Norfolk NR21 8NL
Printed and bound in the UK by TJ Books Limited

The publisher has used its best endeavors to ensure that the URLs for external websites referred to in this book are correct and active at the time of going to press. However, the publisher has no responsibility for the websites and can make no guarantee that a site will remain live or that the content is or will remain appropriate.

Every effort has been made to trace all copyright holders, but if any have been overlooked the publisher will be pleased to include any necessary credits in any subsequent reprint or edition.

For further information on Polity, visit our website:
politybooks.com

Contents

Illustrations   ix
Acknowledgments   xiii
Preface by Emanuele Coccia   xiv

Part 1
First Contacts   1
  Origins   1
  Winters   2
  Weeds   3
  A Scientific Challenge: Remaining or Rising in the Ranks   12
  Customs and Beliefs   22
  Lichen Erotics   34

Part 2
To Describe, Name, Represent   45
  A Challenge to Representation   45
  Music = Mushroom   72
  The Far East, Mosses, and *Wabi-Sabi*   77

Part 3
Ecopoetics: Life Force and Resistance   91
  Ruderal   91
  Rousseauist Walks   92
  Sentinel Species   108
  "Lichens of sunlight and mucus of azure"   112
  "Sbarbarian" Glowworm   116
  Ecological Forewarnings   124
  Fragility, Resistance   132

Contemporary "Poethics"	134
"Insurrection of the Humble"	156
Micro-habitats	166

Part 4
Toward a Symbiotic Way of Thought | 173
   The Politics of Lichen: at the Origins of Symbiosis | 175
   Chimeras, Vampires, and Other Common Monsters | 192
   A "Third Place" | 197
   Cohabitation | 210

Envoi: Sporules | 215

Notes | 220
Index of Names | 255
Index of Lichens | 260

Symbiosis

"Back to nature, then! That means we must add to the exclusively social contract a natural contract of symbiosis and reciprocity in which our relationship to things would set aside mastery and possession in favor of admiring attention, reciprocity, contemplation, and respect. [...] Rights of symbiosis are defined by reciprocity: however much nature gives man, man must give that much back to nature. [...]"

Michel Serres
The Natural Contract

"Like a runaway horse, [the mind] takes a hundred times more trouble for itself [...] and gives birth to so many chimeras and fantastic monsters, one after another [...]"

Michel de Montaigne "On Idleness" *Essais* (1571)

"In our moments of confusion
often I feel the need
to contemplate a lichen.
Bring me a mountain
and I'll show you what I mean."

Hans Magnus Enzensberger "Braille" (1964)

Illustrations

This table lists the black-and-white illustrations in the text, called "figures," followed by their corresponding numbers. The color illustrations in the plate section are called "illustrations," followed by their corresponding numbers.

Figures 5 and 12 come from an art project presented by Nathalie Ravier in 2014 at the School of Art and Design in Orleans. Figures 13–16 and illustrations 15 and 16 are microscopy photographs made in 2017 and 2018 by Pascale Gadon-González in collaboration with the scientific imagery platform of the Center for Applied Electronic Microscopy in Biology at the University of Toulouse III – Paul Sabatier (CMEAB).

Figures

1 Foliose and Usnea Lichens, plates in Johann Jacob Dillenius, *Historia Muscorum*, 1741.   14
2 Diagrams of the symbiotic structure of three types of lichens, with the hyphae of the fungi surrounding the cells of the algae.   17
3 Usnea growing on human skulls, from John Gerard, *The Herball, or Generall historie of plantes*, first published in London in 1597.   28
4 © Pascale Gadon-González, *Signatures*, since 1998, series of photograms, film print, variable dimensions.   29
5 © Nathalie Ravier, "Glossaire des termes couramment utilisés en lichénologie," 2014, impressions on tracing paper, p. 47.   36

6a © Leo Battistelli, *Grayish Pink Skin*, 2010–2011, earthenware and pigment, 123.5 x 123.5 cm.	43
6b © Leo Battistelli, *Sea Blue Lichen*, 2019, ceramic and steel, detail.	43
7a Crustose, foliose, and fruticose lichens.	48
7b Fruticose (*Usnea florida*) and Complex (*Cladonia*).	48
8 A pattern of lirelles of a lichen of the *Graphis* genus (© Vincent Zonca, Rio de Janeiro, 2020).	56
9 Bernard Saby, *Untitled*, 1958, oil on canvas, originally in color, 130 x 97 cm (© Galerie Les Yeux Fertiles, Paris).	69
10a Ogata Kōrin, *Tang Gong Wang*, early 18th century, screen in two panels, originally in color, 166.6 x 180.2 cm (© National Museum of Kyoto, Japan).	89
10b Collections of lichens in the herberia of Geneva's Conservatory and Botanical Gardens, 2018 (© Vincent Zonca).	89
11a © Oscar Furbacken, *Degenerational Crown*, 2020, sculpture in polystyrene, white concrete, iron, and fiberglass, originally in color, Norrtälje, Sweden.	170
11b © Oscar Furbacken, *Beyond Breath*, 2012, series of cut-out photographs (with the painting *Rising* in the background), originally in color, Katarina Church in Stockholm, Sweden.	170
11c © Oscar Furbacken, *Micro-habitat of Rome 2 (Ageless Glow)*, 2020, inkjet print, originally in color, 52 x 80 cm.	171
12 © Nathalie Ravier, "Glossary of terms currently used in lichenology," prints on tracing paper, p. 19.	178
13 © Pascale Gadon-González, *Biomorphose* (4991), 2019, gum bichromate print, with *Xanthoria* lichen, Rome, Italy, 30 x 40 cm.	184
14 © Pascale Gadon-González, *Biomorphose* (5110), 2019, with *Anaptychia* lichen, originally in color, Rome, Italy, pigment print, 30 x 40 cm.	193
15a © Pascale Gadon-González, *Paysage SP*, 2019, gum	

bichromate print, palladium plate, or pigment print,
100 x 160 cm.   207
15b © Pascale Gadon-González, *Cellulaire*, 2018, with *Xanthoria* lichen, gum bichromate print, 38 x 28 cm.   207
16 © Pascale Gadon-González, *Conjonctions 1*, 2018, gum bichromate print, 38 x 28 cm.   208

Plate Section

1 Anton Elfinger, medical illustration (lithograph) of *Lichen ruber pilaris*, in Ferdinand Hebra, *Atlas der Hautkrankheiten* [*Atlas of Skin Diseases*], published by the Imperial Academy of Sciences (Vienna: Braumüller, 1856–1876) (© F. Marin, P. Simon / Bibliothèque Henri Feulard, Hôpital Saint-Louis AP-HP, Paris).
2 *Herpothallon rubrocintum* lichen, Brazil, Botanical Garden of São Paulo, 2018 (© Vincent Zonca).
3 Crustoce and foliose lichens growing indiscriminately on the trunks of bamboos and a plastic chair, Brazil, Ubatuba, 2020 (© Vincent Zonca).
4 Graphidaceae lichen of the *Phaeographina* genus, Colombian Amazon, Ticuan region, Rio Amacayacu, 2018 (© Vincent Zonca).
5 Antoni Pixtot, *Saint George*, 1976, oil on canvas, 194.5 x 97 cm. (© Dalí Theatre-Museum/ © Antoni Pixtot 2022).
6 Foliose lichens grow by rising through, by "licking," the surface of their support: here, a *Xanthoria parietina* on a deciduous tree branch, France, Burgundy, 2019 (© Vincent Zonca).
7 Chen Hongshou, *Plum tree in blossom*, hanging scroll, ink and color on silk, 124.3 x 49.6 cm. (© National Palace Museum, Taipei).
8 Leo Battistelli, *Globe Lichen*, detail, ceramic, 2020, exhibition at the Casa Roberto Marinho in Rio de Janiero, 2020 (© Fernanda Lins).

9   Leo Battistelli, *Red Lichen 1*, 2019, ceramic, steel, and acrylic, exhibition at the Casa Roberto Marinho in Rio de Janiero, 2020 (© Fernanda Lins).
10 © Oscar Furbacken, *Micro-habitat of Stockholm 1*, 2020, photographic print, 70 x 105cm.
11 © Oscar Furbacken, *Urban Lichen*, 2009, Sweden, Stockholm, Swedish Royal Academy of Beaux-Arts, photographic print on painted paper, 500 x 560 cm.
12 Luiz Zerbini, *Mamanguã Reef*, acrylic on canvas, 293 x 417 cm. (© Eduardo Ortega).
13 Yves Chaudouët, *Lichens 1*, lithograph, Clot Studio, 1998 (© ADAGP, Paris and DACS, London 2022).
14 © Pascale Gadon-González, *Cellulaires Contacts 5*, 2017, with *Evernia prunastri* (7296), lichen collected at La Vergne in Charente, 2017–2018, pigment print, 30 x 30 cm.
15 © Pascale Gadon-González, *Bio-indicateur Usnea*, 1999, lichen collected in Corrèze à Maymac, lambda print mounted on Dibond, 120 x 80 cm.
16 © Pascale Gadon-González, *Bio-indicateur Cladonia coccifera*, 1998, lichen collected in Ariège, lambda print mounted on Dibond, 120 x 80 cm.

Acknowledgments

This inquiry, which mixes very different domains, has led to many lucky and generous encounters. I would like to thank all those with whom I've been able to discuss this subject; the book is the result of these exchanges, and those human interactions, invaluable to me.

My thanks, especially, to Emanuele Coccia, as well as to the scientists and artists who have been so kind in opening their studios, their laboratories, and their collections to me, although the lines between these places often blur: Philippe Clerc and the Conservatory and Botanical Gardens of Geneva (Switzerland), Teuvo Ahti and Saara Velmala of the Botanical Museum of the University of Helsinki (Finland), Tomoyuki Katagiri of the Hattori Laboratory of Nichinan and Yoshihito Ohmura of the National Museum of Nature and Science of Tokyo (Japan), Jean Vallade and the Association Française de Lichénologie (in Dijon), Adriano Spielmann (in Brazil) ; Pascale Gadon-González, Yves Chaudouët, Oscar Furbacken, Luiz Zerbini, Leo Battistelli, Nuno Júdice, Olvido García Valdés, Jaime Siles, Claire Second, and Marie Lusson. Deep thanks as well to Nicolas Cerveaux, Ambroise Fontanet, Marjorie Gracieuse, Doriane Bier, Hector Ruiz, Jérôme Bastick, Maxime Chapuis, Ricardo Marques, Patrick Quillier, Mitsuhiro Kishimoto, Chantal Van Haluwyn, Éléa Asselineau, Guillaume Milet, Pierre-Yves Gallard, Emilio Sciarrino, Inès Salas, Makiko Andro, Henry Gil, Clémence Jeannin, Johan Puigdengolas, Miguel Arjona and, above all, to Thérèse Zonca-Philippe, Gabrielle and Camille Philippe.

Preface
Beyond Species

From the beginning, life has seemed to us to be divided into incompatible forms: the dandelion (*Taraxacum ruderalia*) has nothing in common with the squirrel (*Sciurus vulgaris*); the butterfly (*Morpho menelaus didius*) can't be compared to a cork oak (*Quercus robur*). Life immediately appears multiple, dispersed into incompatible and irreconcilable forms. The diversity of species and forms is both obvious and threatening: no sooner is it established than it seems to show both diachronic instability (evolution) and synchronic instability (the struggle between species and competition, ecological disasters).

Yet this multiplicity that divides life in a formal and ontological way is not all that obvious after all. The division of life is always problematic. Plato was the first to understand this, as in one of his best-known myths. In *Protagoras* (320d–322c), he tells how the immortal gods wanted to create forms of mortal life and charged two giants, Prometheus and Epimetheus (literally: the one who thinks of things first and the one who thinks of them later) with the task of granting each species appropriate powers. Epimetheus asked to be allowed to do the distributing, giving some creatures "strength without speed," and others speed with less strength; some were armed with weapons, while others were able to hide, being small, or to protect themselves because of their large size. In the distribution of attributes, Epimetheus sought balance and made extinction impossible for any species. He gave them fur to withstand the cold, hoofs, and tough skin. He appointed different sorts of foods for different species, and arranged an order by which one could eat another, establishing procreative rates accordingly. But he had forgotten one species, humankind: "All the animal species were well off for everything, but man was naked, unshod, and defenseless." So Prometheus stole from Hephaestus and Athena their technical skill, as well as fire (which allowed for the use of this skill), and he gave these to man.

Plato adds three notes to this myth. First, the wisdom of Athena and Hephaestus did not include politics. It was Zeus who later gave this knowledge to men, with Hermes as intermediary, and it was only then that humans were able to make war with other species, "because the art of war is part of the art of politics." The gift of technical skill also established human kinship with the gods. That's why only the human species makes altars and images of the deities. And finally, thanks to this skill, humanity was able to "discover articulate speech and names, and invent houses and clothes and shoes and bedding and get food from the earth." Language is only a consequence of this wisdom and not its foundation. In other words, technical skill precedes reason and is its basis, the condition of the possibility.

The first point to emphasize in this myth is that the war among the species originates in an unjust division. Or rather, what we call biodiversity, the plurality of species, is itself a form of injustice. It is a matter of a division carried out by an imprudent and incapable deity (*Epitheus*) who unequally distributes the characteristic powers of each species. Already there is something extremely radical in this act: species are not characterized according to what they are but according to what they have; any identity is not inherent nature but arbitrary gift. The division of the species is intrinsically arbitrary, resulting from a political form of distribution of what, by nature, belongs to no one. In addition, this division produces a sort of non-species, a sort of proletariat of living beings. For all species, identity is defined by the possession of powers. But humanity, on the contrary, is without possessions, without powers. It is this resentment that defines the war. On the other hand, this resentment prompts a sort of Bovaryism: the desire to be like other species.

Faced with the injustice caused by one god, the myth continues, another god tries to compensate. But his solution produces another, double, injustice: the non-species, the most proletarian of species, receives through theft a quality that belongs only to the divine – the possessors of qualities that other species received in usufruct. This additional gift – skill and fire, considered together as the power that allows materials and reality to be manipulated – permits humanity to establish itself in a relationship of hierarchical superiority. The act that was supposed to correct the inherent injustice in the division of powers produces another, even more radical, injustice.

Plato's myth describes the multiplicity of species as inherently arbitrary: a distribution carried out by a minor god, following criteria not the least bit rational, and with a kind of thoughtlessness or disregard (one of the possible translations for the name *Epimetheus*). But, precisely because it is arbitrary and profoundly unjust, and is then followed by a second injustice – which religion and then politics sanction – this division calls for its own suspension. What is called into question is not so much, or at least not only, human nature, but rather the nature of all species. In opposition to the acquiescence that biology, politics, various theologies, but also and especially ecology has shown in the face of the supposed obviousness of the ontological separation of all species, we must offer less conciliatory discourse. The existence of species is not an ontological fact, it's a practical one. That's why the question of taxonomy – the order that separates one species from another, that defines the reciprocal positions of different living species within the tree, or rather the network, of life – as it's appropriate to speak of it now, following the discovery of horizontal genetic transfer, must become a purely political and no longer a genealogical question.

From this perspective, lichens are players in a new politics of the living: as opposed to the ontological subdivisions of life, they define identity through association, not division. They struggle – throughout their lives – to overcome the division of species. The life that traverses all is always one and the same. That is the only way it becomes possible to commune with beings so far removed in terms of taxonomy and genealogy.

It is this kind of biological communism that Vincent Zonca's book affirms by locating lichens – "living beings situated on the margins, in resistance" – at the very core of the thinking about a new biology. These beings, often described as "leprous," "pustular," "tubercled," which are neither plants nor animals, nor singular, require us to rethink the rules regarding the distribution of identities. They also require us to consider any species as a contingency, which the existence of any individual being – and the encounters to which it gives rise – is destined to go beyond. Life belongs to all the species, and the powers that define each of them must be held in common.

Emanuele Coccia

Part 1
FIRST CONTACTS

"The botanist's magnifying glass is youth recaptured."
Gaston Bachelard, *The Poetics of Space*, 1957

Origins

From the very beginning there was this fascination for strange, unusual words with mysterious meanings and barbaric sounds. Lichen. But also: tundra, wrack, cercus, elytron, dolmen, maelstrom, inlandsis, fjord, permafrost, ubac, adret, axolotl, cortex, pollen. An abecedary of nature, its music marked by the etymology of origins: Greek, Latin, or borrowed from other languages. Those words that, as soon as they are pronounced, create a kind of air current, a moment of suspension. Silex, granite, mitochondria, sphagnum, vraic: lichen. The hardness of the central "ch," the strangeness of the final "en."

*

Victor Hugo is one of the rare francophone poets to have dared elevate lichen to the rank of verse. He rhymed it with a fabled monster of Norwegian origin: the "kraken."[1]

*

It was also a kind of fascination with what is neglected, rejected, denigrated. Adolescence of the accursed poet and lichen, of wild grasses and garden sheds, wandering in search of what lay forgotten off the beaten paths, far from what everyone could see, in search of a territory, a "no man's land." Arborescence of the accursed poet who was trying to construct himself by cutting through fields, gleaning, endlessly branching into a sinuous and ever uncertain verticality.

I made lists of everything, lists of word scraps, backbones for supporting an encyclopedic and solitary imagination as I sought refuge in difference, rarity, the unknown. I investigated the most mysterious pharaohs and dinosaurs. I examined the most repulsive insects, the grubbiest worms; I dissected the blisters on trees in search of inner parasites.

Lichen belonged to the imaginary world of my childhood and adolescence. It inhabited the north faces of the deep forests of my native Burgundy and my lonely dreams. It became conspicuous during those winters that lasted forever, swallowing autumn and stretching into spring, when it was the last visible sign on the bark of the Austrian black pines, in a melancholy landscape of gray and fog, the trees displaying only "their agony of strings," no more leaves or colors – skeletal calligraphy reduced to the elemental.

Winters

If there is a season particularly favorable to lichens, it is truly winter. "Supple physiology allows lichen to shine with life when most other creatures are locked down for winter," writes David George Haskell.[2] While many trees lose their leaves and most of the higher forms of plants disappear, they burst forth in all their colors and extravagant forms. Lichens are the "leaves of winter," wrote Henry David Thoreau. A few pixels forgotten on a canvas.

This is the season that inspired the magnificent descriptions of lichens in Thoreau's *Journal* from the mid-nineteenth century and the northeastern US woodlands – or those of Marcel Proust, Francis Ponge, the haiku poets of Japan. For botanists, this is the fallback – or vexing – solution, for lack of anything better when there's nothing left to study. To the aptly named Malesherbes, Rousseau claimed that "winter has [...] its collections of plants that are specific to it, for learning the mosses and lichens." It is also the season when, during the botanical walks that they so loved, the artists George Sand and John Cage, in the respective company of the botanists Jules Néraud and Guy Nearing, temporarily abandoned plants and mushrooms to let lichens take them by surprise.

*

I stared at those balls of disheveled usnea found on a cherry tree in the family garden. They resembled nothing identifiable, lumps of grass or pale mops of hair so dried out they seemed almost mineral. What were they saying to me?

Lichen is what persists when almost all trace of life has disappeared, in the eternal winter of the poles and the high mountains as well. They become visible, appear, in adversity. Lichen: a critical force?

Weeds

Lichen is familiar to everyone, known to no one. You need only ask those around you: everyone sees, more or less, what the word designates; everyone has already set eyes on those aberrant patches on walls, those strange growths on tree bark. It falls into the order of the "infra-ordinary" to adopt Georges Perec's term; it is "what happens every day and what reappears every day, the banal, the quotidian, the obvious, the common, the ordinary, the background noise, the habitual."[3] But no one has lingered over it or is capable of saying anymore about it: language stops there.

In botanical gardens and parks, no signs ever point out lichens or explain them. A laughable, useless presence on walls, tree trunks, and rocks, even downright repulsive, calling up in the imagination a stain, a kind of eczema or leprosy, pruritus, the idea of an unhealthy, oozing excretion, a disheveled parasite sucking the lifeblood from its host. In 1743, the Chinese poet Qianlong wrote:

> This ravenous parasite which, disdaining the earth whose sap it rejects, will go seeking above it a more abundant and better prepared food: countless filaments, which we might mistake for so many strands of gold, bind it indissolubly to the plants that it devours.[4]

At best it is confused with moss or tree bark in the forests, in cities, with guano on the walls or rubbery debris on our sidewalks.[5] It seems not to have its own identity or to be reduced to a matter of "secretion": it is *what comes out*, what the body rejects, what nature produces that degenerates and proliferates if we are not careful. Something neglected, a deviance.

> In this period I was somehow absent from my body, refusing to support it in any way, to the point of letting this beard grow, like mold, like lichen, which would prove each day to be less and less my style.
>
> Jean Rolin[6]

For the ancient Greek scholar Theophrastus, lichen grew from bark. For others, it is "cliff snot" (the Canadian poet Ken Babstock) or "excrement from earth" (*oussek-el-trab* or *ousseh-el-ard* in Arabic, most likely naming the lichen *Leconora esculenta* with its suggestive brown curves). In natural history, (de)classified first among plants labeled "inferior," it was long treated with disdain and discredited. For Albert the Great, Dominican friar and thirteenth-century philosopher, lichen, located at the bottom of the "vegetable" hierarchy, was the product of putrefaction.

Few bother to worry about the loss of this invisible companion. Because of its size and appearance, it does not have the same charisma as seals, tigers, or orchids. It is part of what, for decades, the scientific community has called "neglected biodiversity." This expression was coined and democratized following marine and terrestrial expeditions by the French National Museum of Natural History in Paris.[7] It points to the fact that a large part of the natural world, which is least well known to the general population, of least interest to the media, and often least studied by the scientific community, is also the part richest in still undiscovered species. It is estimated to represent no less than eighty percent of living species: insects, planktons, mushrooms, lichens, and so on.

In 2016, Emanuele Coccia rebelled against a veritable "metaphysical snobbery," by which the living world and plants in particular had been forgotten by philosophy (at least recently) and condemned to "vegetate."[8] It is possible to be interested in ornamental plants (because of the wholly relative power of beauty) and those considered "useful" (the ones that have nutritional or medicinal "properties"). One might also notice those labeled "superior," higher, to better distinguish them from "inferior" plants, lacking flowers, without lures, subterranean, and lowest on the scale of values: "weeds" and "wild grasses." Hence the dubious honorary chair granted to Jean-Baptiste de Lamarck in 1790, as a snub at the end of that great naturalist's unappreciated career, when he was named the Jardin du Roi's Professor of Natural History of Insects and Worms.

Faceless, mineral and inert in appearance, lichen poses a moral and political problem: it inspires no spontaneous empathy. Like other "lower" beings, it has a hard time accommodating anthropomorphism. Unsurprisingly, scientific studies show that the species for which we feel the most empathy are those closest to us from an evolutionary perspective – and with regard to their appearance.[9] For Emmanuel Levinas, it is through the face and body of the other – "otherwise than being" – that a sense of ethics develops, that it's possible for us to take measure of our humanity.[10] Lichens, insects, planktons: so many organisms that offer nothing to hold our gaze, no spontaneous reflection; so many living beings that we can't "look in the face" to question ourselves (even though the naturalist lexicon for describing them, as is so often the case, relies on metaphors drawn from the human body; see below, p. 10). Sylvain Tesson writes:

> To love is to recognize the value of what one will never be able to know. And not to celebrate one's own reflection in the face of a similar being. To love a Papuan, a child, or a neighbor – nothing could be easier. But a sponge! Lichen! One of those little plants roughed up by the wind! That is truly difficult.[11]

A sense of ethics based on anthropomorphic identification can be a start but cannot be the only vantage point from which to act. And the very species least known to us count among those most in danger. This problem of identification is a real challenge on the level of political action. How to engender an active awareness of the environment without easy recourse to empathy, without the emotional appeal of an imploring look or a heart-breaking cry? How to learn from this quiet, immobile life? Neglected biodiversity: poorly known and about to disappear? Must lichen be condemned to the scholarship of specialists or to our idealization of the marginal and compassion for antiheroes?

*

In Europe and in France, where I'm from and which is the point of departure for my obsession and my intended inquiry, how to explain this paradox of the familiarity and invisibility of lichen? It has no common function or market value. We do not eat it, use it, or sell it; thus we do not see it. An almost total eclipse, except in the eyes of a few initiates. The only books I've found that discuss it are specialized botanical manuals. Until the early twentieth century, these were written in Latin – carrying that distinction of language, knowledge, and power.

> Lichenology has never descended to the level of the lay person for the simple reason that lichens are not, in our country, of

any use, at least since medicine abandoned the doctrine of signatures.[12]

Democratic by its very nature – present and visible almost everywhere, even in cities – it is, at the same time, unpopular.

Someone who takes an interest in lichen, who takes the time and trouble to stop in front of a wall, to circle a tree trunk, to climb a roof, and to approach it close up, is thus seen as eccentric, enlightened, unnerving. Imagine: a passerby who stops in the middle of Paris to examine a vague spot, yellowish and scaly, on the far side of a plane tree! Or who suddenly starts scraping the stones of a historic monument – while the tourists have stepped back far enough to admire it or to take selfies with it! A lichenologist can spend an entire day with the same rock. This is the experience of alienation that the novelist and lichenologist Pierre Gascar recounts in the 1970s, while gathering lichen in the Jura countryside:

> What, this scab, these kinds of aborted mushrooms? And probably poisonous to boot. No creature in the world dares touch it, and haven't you noticed how it makes the trees it grows on wither? Now the vague suspicion that my prowling ways seemed to have planted in these villagers' minds became clear. I would be accused of black magic.[13]

*

Lichen has long been considered a repulsive, parasitic, morbid organism, "the pallid gray of old stone," as Émile Zola wrote, unworthy of interest unless it had some use for industry or survival. For evidence, we can cite the long history of its definition and its classification, as well as the original double meaning of its name, which associates it with a skin disease.

Let us take a glance at Shakespeare:

> ADRIANA [...] Come, I will fasten on this sleeve of thine:/ Thou art an elm, my husband, I a vine,/ Whose weakness married

to thy stronger state/ Makes me with thy strength to communicate./ If aught possess thee from me, it is dross,/ Usurping ivy, brier, or idle moss ["bearded lichen," "usnea"];/ Who all for want of pruning, with intrusion/ Infect thy sap and live on thy confusion.[14]

TAMORA Have I not reason, think you, to look pale?/ These two have ticed me hither to this place,/ A barren detested vale you see it is;/ The trees, though summer, yet forlorn and lean,/ Overcome with moss and baleful mistletoe./ Here never shines the sun; here nothing breeds, [...][15]

In these two scenes, lichen is evoked in a negative way to describe characters or their surroundings. It is viewed as a parasite that maliciously kills the tree on which it lives.

From the beginning (and this is still the case today), in ancient Greek, the word "lichen" (*leikèn*) designated by visual analogy both the organism (or the organisms resembling it) and dartre, a skin disease, "*feu volage*" ["fickle fire"] as it was once called in French, that leaves colored lesions and dry, scaly patches (see Ill. 1):

We gradually see the dermis become infiltrated by embryonic elements, thicken, and turn hard and rough; the papillae hypertrophy and sometimes even group in a way that simulates quite irregular and uneven papules. [...] Soon the skin presents a very singular aspect, characterized by the exaggeration of its natural folds, forming a sort of checkered pattern of more or less wide and regular weave. [...] This is the morbid process to which I give the name *lichenification*.[16]

The image described here by Doctor Louis Brocq is striking. It evokes the *thallus* of the lichen – that is, its exterior tissue that includes no leaf, stem, or root, and which is the *form* we see with the naked eye. We find this "vegetalized" image of the body again in another disease, canker, called "the tree disease."

Portraits by naturalist writers have always struck me with their power and richness. Joris-Karl Huysmans and Zola cannot resist playing with the double meaning of "lichen" to describe the physical and moral decadence of their characters by "vegetalizing" them, by reifying them. These lines are particularly memorable:

> The monk entered.
> He was the most senior member of the monastery, even older than the Father Priest, because he was over eighty-two years old. *Thus, as on the forgotten stump of a very old tree*, lentils, lichens, and burrs were growing on his head.[17]
> There was no end to the train of abominations, it appeared to grow longer and longer. No order was observed, ailments of all kinds were jumbled together; it seemed like the clearing of some inferno where monstrous maladies, the rare and awful cases which provoke a shudder, had been gathered together. [...] Well nigh vanished diseases reappeared; one old woman was affected with leprosy, another was covered with impetiginous lichen *like a tree which has rotted in the shade*.[18]

Conversely, scientific descriptions of lichens have continually used nosographic metaphors to characterize different genera. Lichen is very frequently called "leprous" (there is one family of *Lepraria* – the Latin word *lepra* was used by ancient Christian writers to designate sin and heresy, which attacked and damaged objects), or "pustular," "tubercled."

> Patch of eczema,
> an itch the rock can't scratch
> though the wind's scouring pad
> of grit and sleet brings some relief.
> from "Lichen" by Lorna Crozier[19]

Two motives for these metaphors: crustose and foliose species of lichens seem like so many dermatological manifestations on a surface (see Ill. 2) – the bark of a tree, the outer layer of a wall – while the thalli of fruticose species resemble bristles or shaggy hairs on their supports (see Ill. 15). Thus *Usnea barbata* or bearded usnea is also called *barba de viejo*, Saint Anthony's beard, Jupiter's beard, *barba* or *bigote de las piedras* (in Argentina and Chile), old woman's hair tuft (in Arabic). In Finnish mythology, Tapio, god of the forest, is represented with a beard of lichen and moss eyebrows. Victor Hugo describes lichen as "fur" on trees or a "hairdo" on ruins. If you leaf through botanical manuals, such personifications appear everywhere. From the Latin names and the descriptions a whole imaginary world of skin and hair emerges. Lichen is said to be "wrinkled," "scale-like" ("squamous"), "flesh-colored," "full of warts" (*Icmadophila ericetorum*), "veined," "nippled," as well as "hairy," "bearded," "long-haired," "frizzy," or even "bristly" (*Phaeophyscia hirsuta*), resembling horse hair, "surrounded by hairs," or "ciliated," growing in the "armpits of branches" (*Cladonia uncialis*), or "blackish and hanging like the tail of a horse" (*Usnea jubata*), and so on.[20] For Huysmans, the image becomes fantastic:

> Here the tree appears to him as a living being, standing upside down, its head buried in the hair of its roots, raising its legs in the air. [...] In the joints of the branches, other visions reside, elbows and armpits covered in grey lichen; the very trunk of the tree is scored with incisions which spread out like enormous lips beneath rust-coloured clumps of moss.[21]

Used in the West since antiquity for dyes and medicines, lichen would not be identified in the language or clearly distinguished from other realms until the nineteenth century, the period in which it truly began to make its appearance in scientific discourse and the arts.

*

Eclipse – or explosion? The more closely I approach lichen, the more I see it. Everywhere. In the streets, on the trees, all about us, but also among my contemporaries. When I search through poetry collections from around the world, I discover, turning the pages, first one lichen, then another. I uncover it in the titles of literary reviews (see below, p. 135), among painters, printmakers, photographers, sculptors, before finding it featured prominently in current scientific articles. A simple coincidence? No hermeneutic circle, Proustian madeleine, or adolescent romanticism, but rather a burgeoning, a repeating refrain, a revelation. Lichen, and its proliferation, its politics. What does it want to tell us?

As the object of renewed interest, even rehabilitation, could lichen have particular relevance for today's societies, could this be lichen's "moment"? What is hiding there? What does it hide? This book thus has a dual subject. What does it say about us, about our relationship to nonhuman nature, this little organism flattened against walls? But also, how does lichen come to appear today as a catalyst for our images and fantasies, as a necessity for thinking about the twenty-first century and weaving new relationships?

Beginning from this postulate, I decided to conduct an inquiry, without really knowing where it would lead me, what it would tell me. Over the course of many years, it would have me traveling from France and Switzerland, where I began my research, through the old collections of libraries and herbaria, on to Finland, Sweden, Brazil, and Japan, countries rich with lichen. But to complete my investigation, I had to invent yet another type of travel, a methodology that involved working across the fields of science. This means of access, furthered by the anthropological approach, falls into the line of "diagonal" or "oblique" sciences advocated by Roger Caillois in 1959. It also draws from the methodology of so-called "comparative" literature that brings various contexts and fields of knowledge into dialogue. Even more radically than the philosophy of science, ecocriticism, or anthropology, it involves mixing biology, literary and art criticism, ethnology, and philosophy all together at the same time in order to cultivate curiosity. This is the chosen stance of the undisciplined approach of the essay, the symbiotic functioning of lichen: joyously to blend the disciplines and extract the juices. The cross-disciplinary approach of living beings, and especially of plant life (fungi), even more relevant than "ethnobotanical" science, originated in the nineteenth century and has enjoyed growing success in the last few years. And so, to welcome a diversity of voices, this essay will grant much space to quotations.

The moment has come to give lichen a chance, to give it the right to literary existence by making it the *subject* of reflection, by studying its democratic – and popular – power.

A Scientific Challenge: Remaining or Rising in the Ranks

At the end of 2017, in order to investigate the strange world of lichens and their relationships with human beings, after long searches on the Internet, I decided to explore Geneva's Conservatory and Botanical Gardens on the shore of Lake Geneva. Located in what is now the neighborhood of large international organizations (UNO, OMC), the dark, orange-colored façade of the Conservatory rose before me like a temple consecrated to botany, but also like a place cut off from the world. On either side of the entrance, busts of great Swiss botanists face the intimidated visitor preparing to knock at the monumental door. Philippe Clerc ushered me in. He welcomed me with enthusiasm and kindness; his eyes sparkled as soon as I began to "speak lichen." He could talk forever about his lichenology research – as well as about the species he was currently exploring: the small "yellow heart usnea." He began by leading me down to the basement where he showed me the incredible Genevan herbaria that fill almost eighteen kilometers of shelves (see below, Figures 10a and 10b). Each species has been referenced and carefully wrapped for centuries in old newsprint and folded paper. Many lichens are represented there. Cracking open an envelope reveals a specimen: dried lichen from every corner of the globe, mounted on stone or in fragments, accompanied by its Latin name, neatly handwritten. There are also mysterious red envelopes that contain the "types," the three to five percent of samples used throughout the world as references, a kind of standard for the description of each species. Philippe Clerc then led me to the library, one of the most extensive collections in the world on the subject of mushrooms, mosses, and lichens (that group with the enigmatic name of "cryptogams" – "organisms whose nuptials are secret"). Before me stretched shelves of botanical books, in Latin and other languages, including Russian and Japanese, very old editions with magnificent illustrative plates. But I found very few pages on ethnobotany, the anthropology of lichen, its images and discourse.[22] One needed to know how to read between the lines, to understand the conventions, to learn to decode the scientific jargon, to see how the supposedly objective descriptions betrayed in one way or another the state of the sciences, a way of thinking and seeing, a culture.

*

What exactly, outside of the medical sense (lichen-dartre), what is "lichen" for a botanist?

A living being, a trace, an originary bond. It is that small and strange organism, strange because double: a fungus that is joined to an alga, there where everything begins. The novelist Pierre Gascar writes:

> A structure that is basically branched, a fleshy, non-vascular tissue that has the silkiness of scars: lichen retain something of the fetal. Nevertheless there is no plant more perfectly realized, or presenting a more complex form of organization.[23]

Formerly categorized with plants, lichens left their first marks in the work of the Greek philosopher and botanist Theophrastus (371–287 BCE), where we also find the first *botanical* use of the word. There it refers especially to the liverworts or hepatics, plants that grow notably on the trunks of olive trees and resemble mosses and lichens because of their thalli and the fact that they fasten onto trees and stones.

The word "lichen" in the botanical sense appeared in French in the sixteenth century, in particular in the botanical writings of Rabelais where, as in the ancient Greek, it refers to the lichen used to treat dermatological lichen. But it could also designate species of mosses. For many centuries, lichen was assigned to the family of mosses, or even of algae (by Carl Linnaeus), then to a fuzzy middle ground between algae and fungi. In that period it continued to be considered a secretion from the soil, trees, and rocks. About 125 species were identified. Beginning in 1694 with the great French botanist Joseph Pitton de Tournefort (1656–1708), and especially in his *Institutiones Rei Herbariae*, "lichen" – the organism but also the Greek word – became its own botanical category, distinct, finally, from mosses and hepatics. It was part of the "apetalous flowerless" group. We still have yet to define it by what it has (or what it is) rather than by what it does not have (or what it is not).

Some forty years later, in 1741, the English botanist of German origin, Johann Jacob Dillenius (1684–1747), associated with the taxonomic revolution carried out by Swedish botanist Carl Linnaeus, made the first distinctions within this group (although

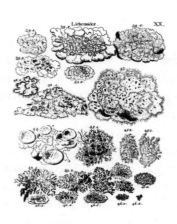 

Figure 1a Foliose Lichens   Figure 1b Usneas
Plates in Johann Jacob Dillenius, *Historia Muscorum*, 1741.

he continued to link them with mosses), beginning with the form of the thallus. In his *Historia Muscorum* (1741), he tried to list all the "inferior plants" (661 species of mosses, algae, hepatics, and lichens) that he knew, with more than eighty magnificent plates that he engraved by hand. Because botanists are very often artists who have learned the art of drawing (and more recently, of photography), when they cannot collect a plant, or in order to note all its details, they draw it. Despite its commercial failure, this book was an astonishing foundational landmark in the history of botanical sciences. The taxonomy of lichens was then significantly expanded and deepened by a student of Linnaeus, the Swedish botanist Erik Acharius (1757–1819). With the development of the science of classification, the number of known lichen species grew from 150 to more than one thousand.

But it was not until the mid-nineteenth century that the use of increasingly sophisticated microscopes revolutionized the sciences (this was the age of vaccines) – and the representation of lichens. They allowed for better identification of them and completely removed all ambiguity by showing their symbiotic nature. Lichens are a combination of two organisms, a fungus and an alga. The German botanist Anton de Bary was one of the first to make that discovery in the 1860s.

In our time, it is molecular biology and computer technology that is revolutionizing lichenology, with the discovery of a third partner and the reclassification of lichen within the fungal kingdom: the realm of *fungi*, mushrooms that also have taken a long time to be defined and classified (half-animal, half-vegetable, they have had their own kingdom only since the 1960s). Phylogenetic research has resulted in more precise classification of species based on the evolution of phenotypes rather than the appearance of thalli: toward the ever more minute.

*

Lichens are neither secretions nor parasites. Gradually, science has shown that the stones, soil, and trees upon which they grow are only supports for affixation; lichens draw no nutritive matter from them.

*

Lichens are very unusual symbiotic organisms. Their thalli are made up of a combination of at least two partners: an ascomycete fungus (ninety percent of the whole), with filaments (hyphae) that contain, at the level of the medulla near the superior cortex, microscopic cells possessing chlorophyll that allow it to feed itself, and very often a unicellular alga or a cyanobacterium. Sometimes another fungus (a basidiomycete yeast) is also present in the outer cortex, the superficial layer of the thallus. This third partner was discovered very recently, in 2016.[24] Thus lichen is sometimes the site of an internal double symbiosis: a fungus–fungus symbiosis in the cortex, within a fungus–alga symbiosis in the whole unit known as lichen.

At present, lichen is considered a fungus (ascomycete or basidiomycete). Moreover, the fungus gives the entire organism its name (*Parmelia sulcata*, for example) and often its external form (in the case of this species, the thallus of the fungus is "a little round furrowed shield"). The fungus thus has the distinctive characteristic of being "lichenized." A very opportunistic organism, it acquired, over the course of its evolution, the ability to enter into symbiosis for nutritional and thus self-serving reasons, and to cohabit with different partners. About forty-five percent of ascomycete fungi have thus been lichenized, that is, paired with an alga. Lichen is "a distinctive lifestyle of fungi," Philippe Clerc told me at Geneva's Conservatory and Botanical Gardens: an alliance, a guild, a marriage. In the same manner, other fungi have been able to combine symbiotically with the roots of trees, shrubs, and herbaceous plants to feed themselves and stimulate the root growth of plants, even in poor soils – these are called mycorrhizal fungi.[25] Alternatively, they may parasitize the flowers, leaves, and stems of certain plants – these are the "rusts" and "smuts."

*

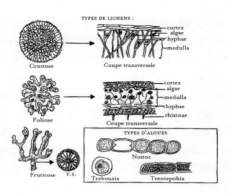
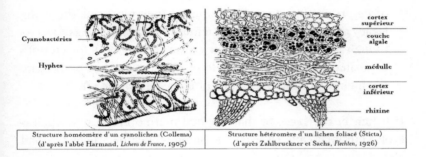

Figures 2a and 2b
Diagrams of the symbiotic structure of three types of lichens, with the hyphae of the fungi surrounding the cells of the algae.

In this sense, lichen is a challenge for science: it is neither a plant nor a vegetable nor a singular "a." It is an adapted fungus – or a double star, its own kingdom.

*

Even today, scientific hypotheses still describe it, along with bacteria, as one of the pioneers in the conquest of terra firma. Because the algae do not use all the products that they produce through photosynthesis and excrete the surplus from their cells, they constitute a kind of manna for these organisms.

At the moment when life first left water, fungi were able to combine with algae in order to better adapt themselves to this new environment and conquer it. According to Pierre Gascar:

> They constitute, on earth, the vestiges of the original aquatic world. [...] The zones for different species. [...] constitute a gradation between the ocean and the heart of the continents. The *Pygmaea* lichen only develops if it is submerged or continually bathed in each tide; a little higher, further inland, the *Caloplaca* is already fine with just sea spray. [...] But even most inland lichens remain fundamentally algae and, in the middle of deserts or on mountain summits, they represent the living legacy of the sea. [...] They are a mummification of algae. [...] As though embalmed under their crusts, locked away for eternity, deposited here by the ebb tide of the epochs, they perpetuate the memory of the most ancient dynasties of the sea.[26]

If lichen bears traces of that original marine life, it nevertheless evolved dramatically. Usneas – the famous "bearded" lichens – had already diversified more than twenty million years ago.

Thus the entire organism, which still lacks a proper scientific name, is a symbiotic association. Each of two (or three) organisms provides for the other what it does not have, in a relationship of close coexistence: the fungus (mycobiont) provides the support (affixation and growth) plus water and minerals to the alga; the alga (photobiont) provides a share of sugars to the fungus thanks to photosynthesis. These exchanges are made with the aid of an organ called the haustorium. Similarly, only in the company of the alga does the fungus synthesize some substances (metabolites) that protect the whole, as, for example, parietin, which gives certain lichens their yellow-orange color (like *Xanthoria parietina*, one of the most widespread in France, whose common names are "yellow scale lichen" [*lichen encroûtant jaune*] or "yellow wall lichen" [*parmélie des murailles*]) to protect them from strong light. It is thanks to such an association that lichen can survive in extreme places: it feeds solely on the ambient air and can dry out while waiting for the next rain.

This astonishing form of organization was not conceived as such until the second half of the nineteenth century. The concept of symbiosis (*sym-biosis*: "to live together") appeared in the 1860s, and was first based on lichen. Gradually biologists discovered numerous symbiotic associations among living beings: coral cannot live without small algae, the clownfish needs the anemone, green plants rely on minuscule bacteria that colonize their cells and help them to sustain themselves by means of light, humans need intestinal bacteria ("micro-biota") – the whole world becomes a great symbiotic equilibrium.

*

It is this atypical way of functioning that has allowed lichen to conquer so many spaces and to establish itself where no plant life can be found. Symbiosis is a factor in ecological endurance. It is estimated that eight percent of the earth's surface is covered with lichen (it constitutes the principal vegetation on the two poles). Lichen is universal.

Lichen develops slowly on a great number of supports – wood, asphalt, gates and fencing, metal and plastic (see Ill. 3), bark, leaves, pine needles, thorns on brambles and ceiba trees, floors, rocks, soil, stones, glass, lava (cooled), and sheets of asbestos – in almost all habitats (except aquatic and cavernicolous). It needs light, air, and humidity. In Rio de Janeiro, I was struck by the omnipresence of lichens on the trunks of tropical vegetation; there are big leafy cakes of them, spongy and multicolored, the size of a hand or larger, that plaster the trunks of coconut palms and frangipanis, creating veritable vertical ecosystems. They are found also – but clearly not only (let us put that romantic fantasy to rest here) – in hostile environments, arid, hot or cold, mountains or deserts, testifying to their powers of resistance. I am thinking of Théodore Monod's *Méharées*, that eternal Sahara so long a fascination for me in my native land. Lichens are more visible there, vegetation being so rare. They even live in the Namib Desert, one of the most arid places in the world, successfully capturing the humidity of its rare fogs. They grow in what are called "intertidal zones," on our coasts where sea levels constantly vary, thus causing variations in salinity, light, and temperature. Italian poet Camillo Sbarbaro writes:

> Lichen thrives everywhere from snow-covered peaks to sea-splashed rocks. It scales summits where no other plant takes root. It is not discouraged by the desert or turned back by the glacier, the tropics, or the poles. It defies the darkness of the cave and ventures into the crater of the volcano. It fears only the proximity of humans. Because of its misanthropy, the city is the only barrier that stops it. If, despite everything, it crosses that barrier, it must seek pure air high in the church towers, or else risk both its health and its particular physiognomy.[27]

As a bioindicator for pollution, it is very sensitive. Here, with echoes of Romanticism and Decadism, Sbarbaro personifies it, makes it heroic, and imagines it as a "misanthropic" organism: in cities, lichen seems to seek refuge only in sites reserved for the church and the sacred, close to the sky. Or perhaps on the roofs of the royal residence at the Tuileries Palace just before its destruction, as the Finnish lichenologist Wilhelm Nylander observed in 1870.

Lichen is one of the last organisms to be encountered near the poles and at high altitudes, at the ice and snow lines (*Xanthoria elegans* has been found in the Himalayas at seven thousand meters). As such, it is a well-known landmark for climbers. In Haute-Savoie, at four thousand meters, some fifty species of lichens are found, as opposed to about ten flowering plants. It can live (the usneas) at temperatures ranging from $-60°$ C to $70°$ C (the alpine lichen *Cladonia foliacea* can photosynthesize at temperatures as low as $-24°$ C). It can utilize the water vapor present in the snow and air. It is the essential link upon which the arctic ecosystem depends. It is the life-form that remains when there is no other life, when the environment has become inhospitable. In the tundra, lichens form a thick carpet that regulates the temperature and humidity of the ground. They provide up to ninety percent of the winter food consumed by Scandinavian reindeer and Canadian caribou. Pierre Gascar writes:

> The slow renewal of the lichen fields where reindeer herds traveled explains why the nomadic shepherds always moved across immense territories. [...] Over the course of his lifetime, no shepherd returned to set up his felt tent in a spot where it had already stood. [...] Nothing ever changed in the landscape; erosion had hardly any effect on ground frozen for eight months of each year. [...] In short, the son relived the life of his dead father and of all those who had preceded him; his entire existence was simply a pilgrimage in the eternal youth of the dead for whom the reviviscence of lichens was the image.[28]

A figure for eternity and survival, lichen is often the first organic matter to establish itself on a support, the first explorer, the first flag planted – and thus the first stage in the food chain. It serves as protection and soft covering for small organisms, especially certain hibernating insects. According to some lichenologists, it is among the pioneer vegetation, playing a central role in the conquest of emergent land, being capable of growing on sand, stones, bare soil, on barely cooled lava flows, where no other plant would yet dare to venture. It retains the dust brought by the wind, accumulating the elements that constitute soil. Crustose lichen becomes embedded in the substrate and, by freeing acids, breaks up rock and pulverizes it, thus contributing over time to the genesis of soil. It then opens the way to a whole dynamic of plant colonization: mosses and ferns, followed by superior plants. The first lichens date back 400–450 million years (the Ordovician period), to just after the appearance of the "Glomales" fungi and the first terrestrial ecosystems.

Lichen: a link to the beginning, a point of affixation.

Customs and Beliefs

Although many species of lichens are too high in acid and prove to be very bitter, certain ones – like the usneas – are edible and are used in traditional recipes or when food is scarce. Soaking them in water reduces their acidity. They have even become a choice ingredient in contemporary "green" gastronomy, among chefs anxious to make a name for themselves: Marc Veyrat of France, for example, serves lichen to accompany cheese; Magnus Nilsson of Sweden combines reindeer lichen (*Cladonia stellaris*), imported from northern Sweden, with fish and eggs. Lichen's food uses remain very limited, however. It primarily serves as a substitute for certain ingredients or as a survival food, in particular for polar inhabitants and explorers (like Sir John Franklin and Sir John Richardson in the early nineteenth century). Species of fruticose lichens are used by the Kirati peoples of Nepal. Reindeer lichen makes up part of the minimal vegetation of the Far North and serves as a staple food for animals there (as it does for the chamois in our own European mountains) and local populations. In fact, reindeer metabolize lichen thanks to an enzyme called "lichenase." Lichen provides them with heat through its fermentation in the rumen, and also with carbohydrates. In Canada at the beginning of the twenty-first century, Native American poets reclaimed their identity in epic songs praising nature and establishing lichen as a cornerstone of their founding mythology. Two women poets of the Innu community, from Pessamit, a First Nations reserve in Quebec on the north shore of the St. Lawrence River, evoke it in this way. Joséphine Bacon (born in 1947) writes about it in two languages:

> You promise me a pure earth
> where you exist
> *Missinaku* lets me drink
> *Papkassik*u runs with me
> lichen feeds me[29]
> moss dries my tears[30]

Natasha Kanapé Fontaine (born in 1991) appropriates and becomes one with the landscape in lyrics that recall Césaire or Neruda:

> Oh my country
> I'll make myself beautiful for the poem
> of my grandmother
> If I were naming you my belly
> if I were naming you my face
> the name of my mountains my river
> the name of my river my sand my lichen
> I'll fix my hair
> like an arctic reindeer
> with the resinous moss of the spruce
> eau-de-vie of the harvests
> [...]
>
> *
>
> My heart throbs
> I slip rings
> onto my fingers
> I put a golden jewel on my head
> tonight, I'll dress
> in my lichen clothes[31]
> I'll fix my hair
> when the drum man comes
> I'll whisper in his ear
> a thousand secrets
> about meteors[32]

This is how she grounds herself at the culmination of her photo-poetic reportage on the remains of her family's ancestral hunting camp at Lake Tetepiskat:

> Walk
> Stand tall

on the moss and the lichen.³³

From island lichen (*Cetraria islandica*) the Lapps make a flour used for baking. Rock tripe (*Umbilicaria esculenta*, or *shi'er* in China, *soegi* in Korea – that is, "rock ear" – and *iwatake* – "rock mushroom" – in Japan) are very popular in Asia where they are eaten as a garnish, after having been collected on the rocks, then dried, boiled, seasoned, and finely chopped (they are represented in a print by Hiroshige II from 1860).³⁴ In India, species of the *Parmelia* genus have been used for curry. The "manna of the desert," actually the lichen *Lecanora esculenta*, constitutes a food source in North African and Asian ecosystems, as Pierre Gascar notes:

> In reality in the history of the Hebrews, the appearance of manna was not a miracle. [...] That Moses' people discovered manna one morning after the evaporation of the dew is explained by the fact that lichens dried by the heat of day and reduced to small strips, scale-like and blending in with the color of the ground, swell with the humidity of the dawn; they are the leaven of the desert.³⁵

The lichen *nęnęndapę* or rather *Dictyonema huaorani*, as it was described and named in 2014, was formerly used by the indigenous Waorani people of the Equadorian Amazon rainforest as an infusion to induce hallucinations during shamanic rituals meant to call down curses.

A Native American recipe for horsehair lichen:³⁶
– Collect the lichens from the branches of trees, using a long stick to reach the highest branches.
– Remove all foreign materials and debris.
– Let the lichens soak overnight in water, preferably running water (river or stream).

Customs and Beliefs

– Make a pit about one meter across and seventy-five centimeters deep, line it with large, round rocks, light a hot fire in the pit, and keep it burning until the rocks are red hot.
– Remove the ashes, put a layer of small branches or wet moss over the rocks, then pile the wet lichens into a layer fifteen to twenty-five centimeters thick.
– Other ingredients can also be added, like wild garlic.
– Cover with other small branches or moss, then with a layer of dirt, leaving a stick planted vertically at the center of the pit.
– When the pit is full, remove the stick and pour water into the resulting hole until the rocks at the bottom can be heard steaming and crackling.
– Then seal the hole with moss and leave the pit like this for about twenty-four hours.
– When the pit is uncovered, the lichens will be a gelatinous mass, two to three centimeters thick.
– This mass can be cut and eaten immediately, or dried in the sun into cakes to be used later.
– The dried cakes must be soaked in water to soften them before using. They are delicious in soups or stews, and can be mixed with other foods such as serviceberries (*Amelanchier*) or cooked with apples, raisins, molasses, or brown sugar.

Horsehair lichen (*Bryoria fremontii*) was an important food source for the earliest inhabitants of the west coast of North America, especially the Salish peoples who called it *wila* (in the Shuswap language) and prepared it in this way. This fruticose lichen grows on conifers and appears in the form of long dark brown hair sometimes extending more than twenty centimeters in length. The old women of the tribe in particular knew how to find the places favorable to its growth and how to distinguish the edible, less bitter species (producing less vulpinic acid) from others that resemble it. First they examined the harvest to determine the color (preferably dark and glossy), then they tasted it: they chewed, reflected, and gave their verdict. As Canadian botanist Trevor Goward (born in 1952) writes, the first North American lichenologists, about one thousand years ago, were these women.[37] The same lichen serves as food to flying squirrels and caribou as well. According to legends of the Salish and Okanagan peoples, it originally came from the hairs of the mythic Coyote, caught in the branches of a pine tree during a fall (in the transcribed account: "he could not untangle his hairs" – that impossibility of untangling evokes for me the symbiotic dimension of lichen).[38]

In the eighteenth century, scientific societies in Europe often sponsored competitions for essays to be written on specific subjects. In 1786, Lyon's Academy of the Sciences, Letters, and Arts gave its prize to Pierre Joseph Amoreux, Georg Franz Hoffman, and Pierre Rémi Willemet for their *Mémoires sur l'utilité des lichens dans la médecine et dans les arts*. Published the following year, the work demonstrated the great variety of uses for lichen, and notably its supposed medical properties.[39] Among those listed are: *Peltigera aphthosa* in milk to treat mouth ulcers in Uppsala, *Usnea barbata* for throat ailments and nasal congestion (it is also used in traditional Chinese medicine), *Usnea hirta* for the skin, *Cetraria islandica* (Iceland lichen) as an astringent for diarrhea or as a cough remedy [...] In his notebooks on "Hygiene," Baudelaire himself reports using it for stubborn colds, with double its quantity of sugar, of course, to counteract its bitterness:

> Fish, cold baths, showers, lichen, lozenges, occasionally; in addition, suppression of everything exciting.
>
>   Iceland lichen   125 grams
>   White sugar   250 grams
>
> Steep the lichen, for twelve to fifteen hours, in a sufficient quantity of cold water, then drain the water. Boil the lichen in two liters of water, on a slow and continuous flame, until the two liters have dwindled to one, remove the scum once, then add the 250 grams of sugar and allow it thicken to

the consistency of syrup. Allow it to cool again. Take a large tablespoonful three times daily, morning, noon, and night. Do not be afraid to increase the dose, if the crises become too frequent.[40]

And here is the fourteenth-century Chinese poet Kouo Yu's treatment for melancholy:

My heart is wounded; the sun sets and two ducks fly off.
For you, I spread and then cooked the lichens [*niu-lo*, medicinal usnea].[41]

*

Certain medical beliefs are linked to analogy. The doctrine of signatures originated in China and began with a conception of humans as reflection and microcosm of the world. It then spread through the rest of Asia and into Europe. In the sixteenth century, the Swiss physician and mystic philosopher Paracelsus (1493–1541) maintained that all natural elements (vegetable, mineral, animal) were signs that formed a (divine) language and corresponded with human anatomy. These analogical beliefs, the idea of a correlation between the ailments of the human organism and the forms of plant organisms especially, can be observed in many cultures.

The form and color of the thallus in certain lichens are suggestive in this way: *Lobaria pulmonaria* or "lungwort," whose ribbed form evokes pulmonary alveoli, was used for respiratory infections like tuberculosis (it is also one of the lichen most sensitive to air pollution!). *Xanthoria parietina* was used to treat jaundice because of its bright yellow color. *Peltigera aphthosa*, because of its tubercular soredia, was a treatment for mouth ulcers. Bearded usneas counteracted hair loss, and *Usnea plicata* and *Parmelia saxatilis*, which grow on human skulls exposed to fresh air, as in the case of hangings, was used for treating epilepsy and hemorrhages, and sold for exorbitant prices.

Figure 3 Usnea growing on human skulls, from John Gerard, *The Herball, or Generall historie of plantes*, first published in London in 1597.

In 1668, the German doctor Michael Ettmüller wrote:

> Usnea [from the human skull] was tested for the treatment of dysentery. Drinking six to twelve grains of usnea from a hanged or dead man is said to have a marvelous effect. [...] There was a fine experiment done with usnea in England, which was related to me by a member of the Royal Society. To treat a severed vein from which flowed very bright red blood, the victim first held the usnea in his hand and immediately the bleeding stopped; as soon as he released the usnea, the blood bubbled out. Then, by taking it in his hand again, he staunched the bleeding, to the great astonishment of the spectators.[42]

Of course in the West, beginning in the seventeenth century, these uses were gradually discredited and viewed as superstitions. In his work on lichens in 1893, Alexandre Acloque denounced the "absurd doctrine of signatures."[43] Conventional modern medicine recognizes only the use of Iceland lichen.

Since 1998, French artist Pascale Gadon-González (born in 1961) has been making photograms of lichens on film, called *Signatures*. Nothing is retained of the lichens but a shadow, an imprint. The title, a reference to Paracelsus' theory, is based on the idea that nature is an enigmatic alphabet (of "fabulous forms"), just like tree branches that grow differently in reaction to any given environment and constitute a trace of it. Lichens are related to writing by this image of the signature: "a repertoire of forms/signs." The (positive) film technique allows for a delicacy that relates these shadows to a form of calligraphy. In Japan especially, image and sign are porous; drawing and lettering are both done in single gestures, ideograms are calligraphed and form images, just as drawn branches can evoke writing.

In the West, one of the first traces of such medical concepts is found in Pliny the Elder's *Natural History*, from the first century CE. The author formed his theory precisely around lichen: the best remedy for the disease called "lichen" could only be the organism

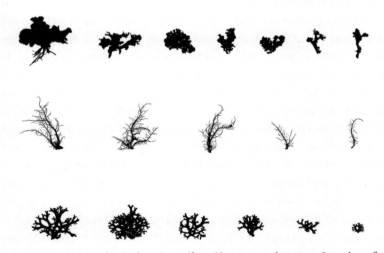

Figure 4 © Pascale Gadon-González, *Signatures*, since 1998, series of photograms, film print, variable dimensions.

with the same name. The visual analogy, which is at the source of the polysemous word in ancient Greek, also involved this therapeutic association:

> But of lichen, which is so disfiguring a disease, I shall amass from all sources a greater number of remedies. [...] Remedies, then, are pounded plantain, cinquefoil, root of asphodel in vinegar. [...] The plant lichen (*Marchantia polymorpha* [which is no longer classified in the genus of lichens; it is a hepatic!]) however is considered a better remedy than all these, a fact which has given the plant its name. [...] This plant removes also marks of scars; it is pounded with honey. There is another kind of lichen (*Lecanora parella* [which is classified today among the lichens]), entirely clinging, as does moss, to rocks; this too is used by itself as a local application.[44]

Remedy for the cursed poet, lichen is also used in perfumery: *Evernia prunastri* (oak moss) – the Guerlain family made it famous with its *Chypre* perfume in 1840 – and *Pseudevernia furfuracea* (tree moss) provide the base note in some perfumes considered "woody." Between six and ten thousand tons of these lichens are still used annually. After storage, which allows for fermentation and hydrolysis, an essential oil with a woody odor is extracted. It is used, for example, in Dior's *Eau sauvage*.

Lichen has even been found at the bottom of a vase in an ancient Egyptian tomb. It was used for embalming. Mixed with sawdust, myrrh, and spices, *Pseudevernia furfuracea*, imported from the Greek isles, was place in the corpse after the organs had been removed. We do not know if the Egyptians understood the aromatic and antibiotic properties of lichen, or if they were using it as a light, absorbent material, but the choice was a good one because it helped to inhibit certain bacteria. Mixed with flour, it was also used to make a kind of bread that was entombed with the mummy to provide a postmortem snack.

*

Customs and Beliefs

Since ancient times, just as Theophrastus and Pliny note, lichen has been used widely as a colorant for painting and for textile dyes (especially for dyeing wool), in order to obtain colors that are usually difficult to produce. Mixed with lime, urine, allowed to ferment with ammonia, or else dried and then pulverized, it provides various shades of reds, blues, and purples (as in *The Lady and the Unicorn* tapestry from 1500), as well as yellows, and has been used for dyeing Scottish tweed, Navajo rugs, and ceremonial dress in Alaska.

Mémoires sur l'utilité des lichens dans la médicine et dans les arts, from 1787, inventories the various shades obtained according to the names used by French dyers.[45] An entire polychromy, a poetry of names explodes before our eyes:

doe color
dark chamois
hazel red
America gray
pale madder
wine dregs
vinegar
tobacco
reddish doe
rotten apricot
light doe
reddish thorn gray
dark sea foam
calf hair
beaver
vicuna gray
rye straw
Levant sand
sour wine
dark nankeen
boxwood
vinegar dregs
hat gray
common
dead leaf
dead vine leaf
doe belly
sulfur
old belly
toad
young belly
toad
viper belly
chamois
well-fed doe belly
ox hair
plum
madder
beaver
crimson bronze
false crimson
reddish brick, and so on.

The famous Scottish fabric Harris Tweed was originally dyed with lichens of the *Parmelia* genus (*Pleurosticta acetarbulum*) steeped in an ammoniacal solution prepared with urine. Until the late nineteenth century and the industrial revolution, the harvesting of orcein (generally species of the *Roccella* genus) – the colorant extracted from lichen – flourished locally in Europe.[46] But the pollution resulting from its extraction as well as its low yield made it difficult to use on an industrial scale.

*

Lichen imposes its slowness. Often growing in extreme environments, the crustose species in particular, lichen grows, on average, only 0.2–1 mm per year, depending on humidity levels. The growth record is ten cm in one year, in contrast to one cm in a millennium for the crustose lichen *Buellia frigada*, which actually manages to live in the arid, frozen region of the McMurdo Dry Valleys in Antarctica. However, because of their long life and slow growth, lichens like *Rhizocarpon geographicum*, for example, can be used in archeology and paleontology for the dating of rocks, ruins, and moraines, a technique called lichenometry. In that regard, lichen is much like the rock on which it adheres: its relationship to time is almost unchanging. The poets Hans Magnus Enzensberger and Guillevic both point this out:

> X
> But as for lichen itself,
> lichen has all the time in the world.
> That ancient one there
> at your feet hardly paid
> attention to Barbarossa[47]
> when his boot crushed it.
> XII
> Our truths are less durable
> lichen covers dead wood
> washes stone idols

survives on churches and ruins.
White is the reindeer cladonia
almost but not completely white.[48]

As the lichens
take care of time.[49]

They emphasize the ethics of slowness that lichen offers us. Lichen allows us to contrast our human finitude with the impression of nature's eternity, very much like the mineral kingdom, great tectonic principles, and astral revolutions.

The still explosions on the rocks,
the lichens, grow
by spreading, gray, concentric shocks.
They have arranged
to meet the rings around the moon, although
within our memories they have not changed.[50]

Common on monuments and tombstones, it is even more frequently compared by poets to what resists history (cenotaphs and funeral urns for Jaime Siles, stele for Lucien Wasselin, mausoleums for Hans Magnus Enzensberger) or to what precedes history (Matthieu Messagier speaks of "lichen-coelacanth").

In our own time, there are lichenic substances (acids) that greatly interest researchers. More than seven hundred molecules have already been identified. Produced through symbiosis, they allow lichens to better adhere to substrates, and to better withstand light radiation, certain pollutants, and temperature variations. The doses are minuscule of course, but they could have many medical and pharmaceutical applications, especially in the fight against cancer.[51]

Lichen Erotics

The Greek verb *leikein* – αείκειν – means "to lick." Was lichen so named because of the speed with which it imbibes or absorbs water, like a tongue or mucous membrane? Lacking roots and growing on substrates that do not retain moisture, it absorbs through the whole thallus the water it receives thanks to the goodwill of the environment (dew, runoff, mist, rain). Is the thallus a tongue set into the bark? Or does lichen "lick" because it gently bonds to the support – bark or rock – with this light caressing motion?

First thing to note: the suffix -ην added to the radical of the Greek verb transforms it into agent; the intrepid lichen is, in fact, *the one that licks*, it is *the licker*.[52] Reserved for the organism – and the disease – is this gesture, an *actio*, a perceptible movement of the tongue. Lichen is the power of touch (see Ill. 6).

Second thing to note: from this primary metaphor of "lichen/licker" emerges a second metaphor; lichen *licks* its support, it *tends* toward it in a movement of desire – but it *is not* its support. This is a movement of closeness, of tangency – not of coincidence.

When we say the word "lichen" in French, the relationship to this action does not strike us. Although the Greek verb evolved phonetically into *lecher* in French (*to lick* in English, *leccare* in Italian), the name that designates the plant itself is still the original ancient Greek word, with its hard pronunciation. This calque tends to mitigate the erotic potential of the word, to harden it into archaism. In Darwin's English language, "lichen" is comparable – only phonetically – to another word: "liken," meaning "to compare," "to assimilate." We no longer lick, but rather we compare, we move closer. As US poet Brenda Hillman (born in 1951) writes, "lichen is a metaphor of the metaphor" or rather "the metaphor is a metaphor of lichen." In both cases, lichen points toward a dialectic of touch and relationship, of presence and absence, of resemblance and desire: *to lick like*.

*

Lichen thus rests on an image that describes a kind of relationship to the world. Its name conveys an exogenous movement, an exit and extension beyond the organism, which is a movement of desire: the desire for contact.

> Lichens are to other plants somewhat like mollusks are to other animal species. They represent the organisms living closest to their supports; with them, life is not separate from adherence. [...] The passionate need of the being to hold onto the world. [...][53]

Pierre Gascar sees in lichen a primary, maternal bond; he links it to the first stages of life when lichen aided in the transition from water to land. From then on, with the strength of its tiny barbed spears, lichen has clung to the world as best it can. Thus many species grow horizontally, adhering to their support, not growing away from it and toward the sky as most plants do. It is as if lichen, unwilling to cut the cord, remains attached to the maternal belly.

*

Umbilicaria is the name of one genus of lichens. Its circular form evokes a navel, with its single point of attachment located at the center of the thallus. The US biologist Robin Wall Kimmerer (born in 1953) writes:

> Where the umbilicus anchors the thallus to the rock, the soft skin is dimpled, with little wrinkles radiating about its center. It looks to all the world like a belly button. Some are such perfect little navels that you want to kiss them, like a little baby belly. [...] The trough can also collect debris, the lichen equivalent of belly-button lint. [...] How fitting that this ancient being, one of the first forms of life on the planet, should be connected to the earth by an umbilicus.[54]

*

However, unlike with plants, this singular and original point of affixation is not – or is no longer – a root. And in fact, it can be more or less anchored and firm. Crustose lichens become embedded in rock through the action of powerful acids and do not have only one point of attachment. Lacking an inferior cortex, the organism as a whole is fastened to the support and thus much closer to it. In this sense, the support in some way becomes an integral part of its body. Other types of lichen, like foliose and fruticose, become attached at a fixed point by what are called hapteres (a sort of "crampon," from the Greek *haptein*, "to attach") or rhizines ("little roots," made of filaments called hyphae), like algae, and are easily detachable with bare hands.

> Other [types] live on the sand. Thus, given the impossibility of clinging to an element that is continually giving way beneath them, tossed by the wind that sometimes lifts them off the ground, caught in the sun wherever they go, they learn to be self-sufficient. They curl up and clump together. These are the *deracinated*, the lichens without fixed homes, like the *Parmelia vagans* of the Kazakh Steppes.
>
> Camillo Sbarbaro[55]

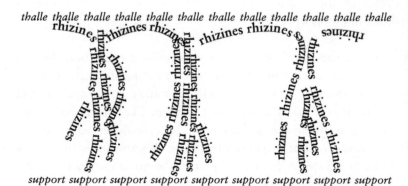

Figure 5 © Nathalie Ravier, "Glossaire des termes couramment utilisés en lichénologie," 2014, impressions on tracing paper, p. 47.

Emanuele Coccia writes, on the subject of plants: "Their absence of movement is nothing but the reverse of their complete adhesion to what happens to them and their environment. One cannot separate the plant – *neither physically nor metaphysically* – from the world that accommodates it."[56] Lichen has just as intimate a relationship with the world, but it favors a tenuous, partial, minimal adhesion. In this, lichens are like *epi*-phytic plants that remain *on*, on the surface of their supports. If roots and penetration are minimal, the withdrawal of nutritive matter is impossible (lichens do not parasitize their supports). Therein lies all the ambivalence of lichen: a singular and sensual relationship with the support that accommodates it, but which is only a support (although also a vector of humidity); meanwhile the organism is entirely turned in another direction, toward the air (which provides its nutritive matter).

*

There exists a genus of lichen that grows on tree trunks and walls sheltered from strong rain, and whose name conjures images of skin and disease: the *Lepraria* or leprose. Their thallus is powdery, composed of tiny granules, and easily detached from the substrate. The terms "leprosy," *Lepraria*, and *Lepidoptera* (butterflies) all come from the same Greek word, *lepein*, which means "to peel," "to remove the bark." Lichen is thus a peel, a desquamation, a second skin. The *Umbilicaria* has that look of desquamations, peelings, off rocks; it gives the impression that the slightest wind will release it into the air, its cord cut.

*

Let us return to the metaphor of the licker. It seems to me that we can understand it in two ways. The action of licking can describe not only lichen's tenuous adhesion, as though set there on rock or tree trunk, grazing the surface, suspended by all the delicacy of its rhizines or hapteres; but also the movement of growth for certain lichens, which cover and hug their support as they develop. In this sense, to return to another metaphor, lichen is a kind of clothing, a tongue lapping skin. In Chinese moreover, lichen is called 地衣 [*di yi*⁻], "clothing of the earth." We speak of fire "licking" a body or an object in the same way: an airy touch, contact without contact. Lichen: a painting hung on the world, possibly just barely penetrating the wall, held (back) by a single, slender point of attachment: the rhizine/nail.

*

In its quest for contact, lichen lightly touches its support. The thallus, like a tongue, lip, or mucous membrane, imbibes, gorges itself on water. Lichen is a sensual – and sexual – organism.

> Robinson hesitated for many days on the threshold of what he would later call the *plant way*. He returned to circle around the quillai tree with suspicious looks, finally finding a way to enter the branches that spread under the grass like two enormous black thighs. In the end he stretched out naked on the lightning-struck tree. [...] He spent long months in a happy liaison with Quillai.
>
> Michel Tournier[57]

Like moss, lichen harbors true erotic power, despite the general disdain it has long inspired. Earlier, we saw that the thallus is like hair or fur. It hides and reveals the crotch of branches, the inert skin of a tree or wall. Usnea is the beard of the sage, but also the naked pubes. In humid weather, foliose lichens quiver like the lips of the vulva. In the West since antiquity, the cave (before which Diana bathed in Ovid or posed in Cranach's painting) and mossy trees, sites familiar to nymphs or Pan, evoke mysteries that writers love to explore.[58]

Canadian novelist Alice Munro (born in 1931), who won the Nobel Prize in 2013, develops this image in her short story collection, *The Progress of Love* (1986). In the story aptly entitled "Lichen," she tells of the day that David spends each year at the home of his former wife Stella near Lake Huron. On this particular occasion, he introduces Stella to his latest conquest Catherine, but he also admits to Stella that he is growing tired of Catherine and has taken an interest in a younger woman, a certain Dina. Of Dina, we will see only a photograph, the one David shows Stella over the course of the story. But in fact, it is less a portrait than a trophy, a fragment, centered on Dina's crotch. We find here Gustave Courbet's "The Origin of the World" (1866): the body presented as if head and arms had been amputated. The photograph appears at three key moments in the story that it weaves through and punctuates: at the beginning, middle, and end, and each time in connection with the repudiated wife. The progressive aging of the photo – and of what it shows – makes it an indicator of the passing time, a kind of conceit.

> David has one hand in the inner pocket of his jacket. He brings out something he keeps cupped in his palm, shows it to Ron with a deprecating smile. "One of my interests," he says.[59]

With its first appearance at the beginning of the story, the image is left a mystery, simply summarized by the narration. David, who shows the photo, does not describe it. All the narrator describes is the gesture of sharing it. The reader is left in the dark, waiting. Nonetheless, the erotic content is anticipated by the gesture – "he keeps cupped" – like Venus rising from her shell, the crotch very much the place of intimacy and eros, "an impression of intimacy" as Bachelard would say. And yet, Dina is presented straightaway in a negative, reified fashion by the man, as a thing among other things ("one of my interests"). Later, the image is revealed to the reader, through the eyes of the ex-wife, a country woman who exposes the predatory behavior David flaunts even while allowing a glimpse of her jealousy.

> Stella puts the paring knife down and squints obediently. There is a flattened-out breast far away on the horizon. And the legs

spreading into the foreground. The legs are spread wide – smooth, golden, monumental: fallen columns. Between them is the dark blot she called moss, or lichen. But it's really more like the dark pelt of an animal, with the head and tail and feet chopped off. Dark silky pelt of some unlucky rodent.[60]

David's new prey is sprawled there, provocative, free for the taking, like a painting of ruins by Hubert Robert or Giorgio de Chirico, or like an installation by Marcel Duchamp (*Étant donnés*, 1946–1966, which presents the body of a woman without a face, visible through the hole in a wall, and questions the inherent voyeurism in the concept of an exhibition). The legs in the foreground are described as fallen columns and they direct the gaze toward the pubic area, then toward the breast in the background. The crotch appears as a blur, hence the attempts to capture it with successive, nonhuman comparisons: the hairy pubes seems first like moss or lichen, then like a pelt, the pelt of a small animal without perceptible body parts.

And now, look, her words have come true. The outline of the breast has disappeared. You would never know that the legs were legs. The black has turned to gray, to the soft, dry color of a plant mysteriously nourished on the rocks.[61]

At the end of the story, after David's departure, the photo reappears and is briefly described by Stella. The image has been altered by time and has transformed. The author emphasizes the mystery: the comparisons to columns of statues and to moss and lichen, which were made initially, have come true (Dina has become mineral, the pubes a plant growing on rocks, as if Stella's words proved to be performative). Thus it is a story of a metamorphosis which, unlike Ovid's, is transferred – and thereby fixed – onto an image, an icon. At the end of the story, Dina's body is naturalized, petrified, and her pubes lichenized. Like a Courbet coupled with an Arcimboldo, a Kahlo, or a Dalí. "The Origin of the World" to which the photo refers no longer portrays the matrix of the feminine sex, but primitive lichen: woman as lichen. Dina's photo, left behind by David in the curtains of Stella's living room, shows her as the new repudiated woman, abandoned and left lying fallow, like lichen.

As in her other short stories, Munro reflects here on female heroines, present or represented, facing the male gaze with its predatory potential, and that of other women. Marked by the Ontario countryside where she lives a quiet life, Stella, like the author, is also nurtured by the natural elements surrounding her.

The resemblance of lichen to pubic hair, combined with that of their substrate to skin, finds its ideal expression in photography, as Alice Munro demonstrates in her short story. Various artists have developed this imagery of tuft and bush (that is the meaning of the word "fruticose" in Latin) on the human body.

French photographer and writer Paul-Armand Gette (born in 1927) had developed a photographic universe in which his favorite themes are femininity and nature (that is, the female model and the landscape). His texts and photographs often combine them in an erotic context (spring and fountain, moss and lichen, flowers and tree trunks, pebbles) and are filled with mythological references to nymphs and Artemis-Diana, echoing a long pictorial tradition (the smooth-cheeked nymphs of Botticelli, Cranach, Lotto) and the mischievous eroticism of poet Francis Ponge.

> For him, there is nothing marmoreal about mythology; it is incarnate. What interests him remains the pink flesh tones of a crushed strawberry on very white skin in close proximity to the fleece that is more pleasing than its golden counterpart.
> Jean-Paul Gavard-Perret[62]

Like lichen, or the pubes, the artist has "a pronounced fondness for moisture" (whatever flows, oozes, overflows, "spurts") and makes photography a space for erotic exploration. The erotics of the image, as with the painter Bernard Legay (see below, p. 122): the photograph becomes skin, it is made to be touched, licked.

> TL Once you told me that you licked your drawings.
> PA When it's necessary, yes.
> TL And their sources?
> PA As well [...]
> TL But it's an image that you're licking.
> PA Yes, just to show you.[63]

The female genitals – the mons veneris – become a form to play with, a double for the desire of flowers, bark, or moss, a game played as well by Finnish artist Tuula Närhinen (born in 1967) in her photographs, like *Coiffeuse ou la Toilette intime* or *Birch Forest*.

This provocative formal research in photography around the analogy of the human body and elements of the forest (branches, tree trunks, lichens, mosses), and the contrasts in colors and textures that can spring from it are also at the heart of the work of Chinese photographer and poet Ren Hang (1987–2017). Various shots expose the naked bodies of women and men who touch, caress, encircle, hug, imitate, or lick trees or rocks. Taken at night with a flash, they make the pop colors and textures of bodies and surfaces explode, along with the leaves, moss, and lichens sometimes found there. Nature appears clearly, and not without humor, as the metaphor and extension of a human body with which the photographed subject enters into a relationship: maternal lair, intimate cavity, nipple or crotch, everything is suggested while the models – sometimes the photographer himself – call as witness or provoke the spectator with their impassive gaze.

> The lichen pulses under the spring's flow, at least at the spot where the thallus, as long and wide as a hand, becomes a little detached from the tufa and reminds one of the lip of a vulva. Entirely carpeting the small throat out of which the water runs, and letting one think, because of the pulsation that lifts

the edge of its thallus, that it determines more than submits to the slight jerks of flow, the lichen gives the spring a visceral appearance [...], thus endowing the earth with genitals or a mouth, around which to organize itself.

Pierre Gasgar[64]

*

Tongue, erotic tension, bodily sign, lichen also possesses this other erotic potential: symbiotic in nature, its existence is based on the marriage of two (or more) organisms, their exchanges being ensured thanks to an organ called a "sucker" (*haustorium* is the less explicit Latin name, although it means "vase for drawing (water)"). Thus, lichen's symbiosis becomes a long, soft French kiss.

Figure 6a © Leo Battistelli, *Grayish Pink Skin*, 2010–2011, earthenware and pigment, 123.5 x 123.5 cm.

Figure 6b © Leo Battistelli, *Sea Blue Lichen*, 2019, ceramic and steel, detail.

Part 2
TO DESCRIBE, NAME, REPRESENT

> "As their structures and colors are so different from those of most terrestrial plants, I often see them as windows onto another way of viewing the world."
> Oscar Furbacken[1]

A Challenge to Representation

Colors and Forms
At present, about twenty thousand species of lichens have been inventoried, which corresponds to twenty percent of the known fungi species.

> One could say that it constitutes, in whole or in part, the most disparate of substances: starch and flour; wool and murex; gold, sulfur and sealing wax; sponge, cork, and anthracite, parchment and gutta-percha.
> Camillo Sbarbaro[2]

A challenge to classification, but also a challenge to representation. Frizzy, curly, frayed, leafy, or bearded; tuberous or warty, leprous or elephant-skinned, lace or cup; granular crust, almost invisible patch on bark or stone, creepers falling from branches, small projections or mini-escutcheons, coral, tentacles, antennae, tubes or scales, powder or ash: lichen is polymorphous, it a laboratory of forms. What jumps out even more as soon as you look through a magnifying glass: the crevices and valleys of the thallus, the turrets of the apothecia and sharp teeth of the rhizines hidden under the thallus. It is in itself a landscape, an invitation for metaphor and verbal creativity, but also an invitation to touch.

The names (the Latin ones especially) and descriptions by botanists, often poetic and magnificent, are the mark and scar of this; they clearly demonstrate the challenge presented to language. To identify lichens, scientists have shown remarkable sensitivity and poetic inventiveness by approaching them through metaphors and, quite readily, through personification in terms of the human body (skin, hair). Thus there exists a "decayed molar" lichen (*Pertusaria pertusa*, whose apothecia resemble bad teeth or tiny skulls), and well or badly "combed" species (*Cladonia arbuscula* or *Cladonia portentosa*). Scientific works are packed with particularly elaborate images:

> LICHEN of Greece (exotic botanical), species of lichen used for red dyes. [...] It grows in grayish bunches, two or three inches long, divided into small sprigs almost as fine as hair, and grouped into two or three small horns, released at birth rounded and stiff, but subsequently almost as thick as a line, bent into a sickle, and sometimes terminating in two points. These small horns are lined lengthwise with a row of cups whiter than the rest, a half-line in diameter, held up by small warts, similar to the cups of sea polyps; the whole plant is solid, white, and salty tasting.[3]

> *Lichen carpineus* [*Lecanora carpinea*] . [...] Laden with rough patches similar to those on a hand toughened by work.

> *Lichen miniatus* [*Lecanora miniata*]. Foliose, hunchbacked, punctuated, tawny underneath. On the Alps. It forms leaves singly or in threes, hard, ashen, raised and concave, in the

form of a saucer or irregular seashell, punctuated above, a bit yellowish or reddish underneath.

Lichen jubatus [*Usnea jubata*]. Filamentous, hanging; axils compressed. In forests and on rocks in Europe. Blackish beard hanging from trees, like the tail of a horse, which has earned it the name of *jubatus* which expresses this characteristic.[4]

Persoon Opegrapha. Off-white, slightly smooth, uneven crust; deep-set conceptacles, initially oblong, with furrowed disc, then rough, flexuous, creased, deformed, nearly contiguous, with irregular, half-open disc. Acharius describes two varieties of it: one of them, *aporea*, with a leprous, pulverulent crust, and twisting conceptacles opening irregularly.[5]

Scientists have identified four main appearances, according to the form of the thallus. Generally, these are: the *gelatinous* lichens, with the filaments of the fungus branching into the gelatinous mass of the algae, which is the only case where the algae gives the lichen its exterior form; the *powdery or leprous* lichens, which appear in the form of a powder and have no superior surface; the *crustose* lichens, in which the green cells of the algae are confined within the compact tissue of the fungus, the lichen spreading out into a crust inseparable from the substrate which serves as inferior surface; the *foliose* lichens, which form strips with wavy edges, like tiny lettuce leaves, that detach slightly from the substrate even while adhering strongly to it; and the *fruticose* lichens, which resemble a miniature shrub or bush – *fruticulus* – that is only attached by its base to the substrate, and thus easily detachable. In all these cases, only one of the two partners is immediately visible to the naked eye; the other is invisible, hidden within the tissue.

Lichen is not just a flat, horizontal form; it has a third dimension. Its reproductive organs – sexual, the *apothecium* (in the form of small cups originating from the spores), or asexual, the *icidium* (excrescence whose name means "coral" in Greek)

– constitute a whole repertoire of erect forms rising, in proportion to their size, toward the sky. Thus the *Cladonia* have a complex thallus (see Ill. 16): a primary crustose thallus on which grows a kind of foot, point, horn, or tentacle (the *podetium*) surmounted by an apotheceum, sometimes of a different color.

This organism gives its name to a color: *lichen green*, a green tending toward pale and gray, which is no doubt one of its most common and certainly one of its most remarkable colors. At the end of the nineteenth century, this color appeared on color charts and was part of the nomenclature fashionable at the time, along with "moss green," "frog green," "lizard green," "juniper green," "Alpen glacier green," "angelic green," "peacock green, "acanthus green," "fucus green," and even "caterpillar green." According to old botanical manuals, lichen was "rarely a beautiful green," but in the 2010s, this "lichen green" became a color of choice for interior

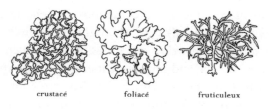

Figure 7a (crustose; foliose; fruticose)

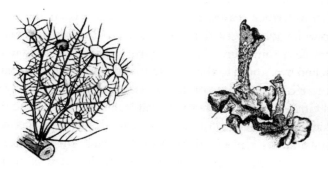

Figure 7b Fruticose (*Usnea florida*) and Complex (*Cladonia*)]

decoration, ubiquitous on the walls of our apartments, in stores and restaurants. Here are its color codes:

RGB 17, 150, 113
HSL 163, 80, 33
HSB 163, 89, 59
CMYK 89, 0, 25, 41

There is an intense polychromy to lichen. Contrary to what we may think, there is not only this greenish or grayish green color, coming from the algae and their chloroplasts, this dull tone so popular today. Lichen also has its flashy or "pop" side. Such shades explode on the palette, intensified by moisture; like tattoo art engraved outside, there are so many individual flourishes, cut jewels, decorating the fingers of trees and the skin of rocks. There are also grays, blues, purples (because of cyanobacteria), "orange and sulfur colors" wrote novelist Joris-Karl Huysmans, to evoke, of course, the shades of the famous *Xanthoria parietina* found in our cities (see Ill. 6), candle or egg yellow as indicated by the name *Candelariella vitellina* (from the Latin *candela*, "candle," evoking the color of the wax formerly used), due to the substance it secretes as protection from the sun. Lichen is a complex play of colors, as mycologist Richard Bernaer (born in 1948) describes in his poetic botanical blog:

> The *Caloplaca*: beautiful plates, beautiful wide, flat surfaces, are often pure marvels of warm tones over warm tones, like our *Caloplaca aurantia* [...], a mosaic of saturated brown-orange lecanorine apothecia, a yellow thalline rim, over the vitelline yellow-orange of the thallus.[6]

I'm thinking as well of the magnificent blood reds of the *Herpothallon rubrocintum* that have accompanied me throughout my time in Brazil, the "Christmas lichen": bright red patches, as though left by bullets – or stars, or dusk – that seem to flow from these wounds as they grow vertically on tree trunks in the luxuriance of the tropics (see Ill. 2). And when the weather turns wet, the colors intensify and come to life, saturating the landscape.

Lichen creates striking compositions according to the colors of the bark and stones it grows on, or the moss it very often lives with. On trees, it can make us think of leprous and aging skin or, alternatively, bring the bark to life with vibrant colors. Videographer Claire Second (born in 1989), who has made a documentary on lichens in Ardèche (*L'Algue et le Champignon*, 2016), described to me her excitement over their explosive colors on the deep black of basal rocks. Barry Lopez observed the same thing in the Far North:

> At first it seems that, except for a brief two weeks in autumn, the Arctic is without color. Its land colors are the color of deserts, the ochers and siennas of stratified soils, the gray-greens of sparse plant life on bare soil. On closer inspection, however, the monotonic rock of the polar desert is seen to harbor the myriad greens, reds, yellow, and oranges of lichens. [...] Occasionally there is brilliant coloring – as with wildflowers in summer, or a hillside of willow and bearberry in the fall; [...] [7]

Italian poet Camillo Sbarbaro offers us an extended pictorial metaphor, emphasizing all the dark tones of lichens and recalling *sfumato*, the "smoke-like" effect of blended light and shadow in the paintings of Leonardo da Vinci:

> Lichen is the most polychromatic of plants. Its color range, which extends from milky white to stygian black, ascends toward all the sharp notes through an orchestration of tones and shades in which the most lavish repertoire of colors is deployed. [...]
>
> Thus, not to leave the limbo of blacks, [we] can yet distinguish [...] a bat black (*vespertilio*), a crow black (*coracinus*), a smoked

black (*infumatus*), a funeral black (*pullatus*), a roasted black (*torridus*), a burned black (*deustus*), an anthracite black (*anthracinus*), a soot black (*fuligineux*), a gloom black (*tenebricus*), an infernal black (*stygius*) . [...] [8]

I recall as well my fascination with the lichen *Hypotrachyna laevigata*, found near Bogota, in the heart of those astonishingly moist mountain landscapes in Colombia, called *páramos*, which are perfect for fog and the development of lichens. Its very light whitish-gray thallus is rimmed with a very intense black, which is also the color of the reverse side of the thalllus, conferring upon it an intimidating gravity.

Moreover, numerous chemical reactions can be used to identify the various species: on contact, the thallus reacts through colors just as vivid and glistening. Also, because of its tinctorial properties, lichen was one of the first known resources for the fabrication of colors.

Natural Images

> It is somewhat of a lichen day. The bright-yellow sulphur lichens on the walls of the Walden road look novel, as if I had not seen them for a long time. Do they not require cold as much as moisture to enliven them? What surprising forms and colors! [...] How naturally they adorn our works of art! See where the farmer has set up his post-and-rail fences along the road. The sulphur lichen has, as it were, at once leaped to occupy the northern side of each post, as in towns handbills are pasted on all surfaces, and the rails are more or less gilded with them as if it had rained gilt. The handbill which nature affixes to the north side of posts and trees and other surfaces. And there are the various shades of green and gray beside.[9]

In this passage drawn from his *Journal*, Thoreau momentarily blurs the boundary between nature and culture. He compares lichen to a human creation, first for the way it "adorns our works of art" (a stone wall), and then by gradually likening it to a handbill. Through its complex forms, lichen invites us to reflect on the imaginary, and on what constitutes *image*, what constitutes *art*.

It becomes part of the human attempt to see images and figures in nature, in the semi-abstract forms of the world: signs or works of art.

> But lichens are beautiful too, adorning rocks and trees lavishly with their chaste embroideries. The lobes of a Parmelia may branch as exquisitely as do the ornaments on a marble temple. [...] The fruiting tips of one Cladonia may flaunt a vermilio purer than the garden's most treasured flower, while the goblets of another might serve as models for the acme of the potter's art. These forms stem from the roots of creation, primal and pure.
>
> Guy Nearing[10]

Science talks about "pareidolia": vague visual or auditory stimuli can trigger the human brain to find meaningful images in forms or objects, often from daily life and on a human scale. These may be butterfly wings, the forms of stones or veins in marble, petals of flowers – a fascination shared by George Sand and Roger Caillois.[11] Nature's power over the imagination leads to analogies (comparisons, metaphors). It often testifies to an aesthetic appreciation based on the idea of a (technical and aesthetic) perfection linked to a religious heritage. Sometimes, it is simply a matter of play or exercising the imagination, creative protocol dependent on mundane accidents of nature. Transcendence versus literality.

With regard to this subject, Roger Caillois speaks of "natural images." Stones, with their suggestive forms and features offering so many possible designs, inspire reverie in the West as in the Far East, where stones with the most astonishing patterns are collected with great interest: *gongshi* and *suiseki*. These "natural images" stimulate the imagination and delight artists. In seventeenth-century Europe, the fashion was to embellish stones. Flemish artists in particular liked to paint images on agate, flint, marble, and alabaster with suggestive forms "whose very texture constituted a landscape" (Jurgis Baltrušaitis) or semi-abstraction.[12] That is also the case with Tuscany's famous *paesine*, stones with marbling that depicts landscapes, and which are treasured items for cabinets of curiosities. The scholar and occultist Athanasius Kircher depicted the pictorial and mystical power of nature in his *Mundus subterraneus* (1664). One of its chapters concerns mineralogy. These natural paintings were dear to the surrealists. André Breton praises them in his essay "Langue des pierres" ["Language of Stones"] (1957).[13] Roger Caillois "read" stones and minerals in the 1960s (*Pierres*, 1966). In the 1860s and 1870s, toward the end of her life, George Sand took up "aquarelles à l'écrasage," what she called "dendrites" (from *dendron*, Greek for "tree"), named for those minerals that, due to solidification, present lines in the forms of arborescences. Watercolor was "dripped" onto drawing paper, then pressed down with heavier paper or cardboard. The forms thus appearing on the paper resembled branches – the branching of trees, hence the name – and were extended or reworked with a brush.[14] "This crushing produces veining that is sometimes strange and interesting [...] with the help of my imagination, I see in it woods, forests, or lakes, and I accentuate the vague forms produced by chance."[15] Nature thus generates aleatory forms for the creative mind; it frees the imagination and the unconscious (André Breton's internal, savage eye). Such processes were praised by the surrealists who developed various techniques, like Max Ernst (1891–1976), for example, with his decalcomania and frottage (in 1925): the artist rubs pencil lead over paper pressed against any surface, a wooden floor, wall, or other texture, in order to make more or less abstract figures appear and let the interior world speak. Lichen: a Rorschach test, a projection screen for desire? All the more so as their vertical presentation – lichen, but also old handbills, spots appearing on walls and tree trunks – creates a "painting effect." In this famous passage, praised by

André Breton in his 1933 essay, "Le Message automatique" ["The Automatic Message"], Leonardo da Vinci links spots on wall, and thus perhaps lichen, to the origins of painting:

> If you cast your glance on any walls dirty with such stains [*macchie*] or walls made up of multicolored rocks, with the idea of inventing some scenes, you will be able to discover them there in diverse forms, in diverse landscapes, adorned with mountains, rivers, rocks, trees, extensive plains, valleys, and hills. You can even see different battle scenes and figures in costume making lively gestures and faces with strange expressions and myriad things which you can transform into a complete and proper form.[16]

Vasari noted a similar anecdote regarding another Italian painter from the same period, Piero di Cosimo:

> He would sometimes stop to gaze at a wall against which sick people had been for a long time discharging their spittle, and from this he would picture to himself battles of horsemen, and the most fantastic cities and widest landscapes that were ever seen; and he did the same with the clouds in the sky.[17]

Like spots and clouds, lichens are powerful generators of forms and revealers of desires.

*

Saxicolous lichens (which live on rocks, boulders, and cliffs) are in themselves veritable sculptors. Although they seem fragile and laughable, these tiny organisms have the power to model their mineral substrates. They act on them extremely slowly; their fungal filaments and acids succeed in fracturing rocky structures and gradually crushing the rock, altering its mineral forms just as water and wind do.

*

Moreover, the patterns traced by certain species, spread over a horizontal or vertical support, have made lichens a favorite metaphor for writing in the eyes of scientists as well as poets. They can evoke anew the mysteries of the sign and of a transcendent natural world (to decipher, contemplate, dream about) or, alternatively, the literal materiality of the drawn form. In Sweden, Tomas Tranströmer studied "The roofing tiles with lichens' script written in an unknown tongue."[18] In 1966, in Caillois' "diagonal" meditation entitled *Pierres* [Stones] that blends science, poetry, and magic, he describes the minerals in his collection as so many enigmatic signs and graphs to read, called "ink lichens."[19] Four years later, in another collection of mineral readings (*Sky Stones*), Pablo Neruda draws out the metaphor of lichen as natural (and original) alphabet by comparing it to a "hieroglyphic" which, etymologically, designates script that is impossible to decipher:

> Lichen on stone: the web
> of green rubber
> weaves an old hieroglyphic,
> unfolding the script
> of the sea
> on the curve of a boulder.
> The sun reads it. The mollusk devours it.
> Fish slither
> on stone, with a bristling of hackles.
> An alphabet moves in the silence,
> printing its drowned incunabula
> on the naked flank of the beaches.
>
> The lichens
> climb higher, plaiting and braiding, piling
> their nap in the caverns of ocean and air, coming
> and going, until nothing may dance but the wave
> and nothing persist but the wind.[20]

I am thinking, here, of the magnificent family of *Graphidaceae*, the favorite lichen of writers, and so mine as well. The best-known species is the *Graphis scripta*, also called "script lichen" or "secret writing lichen," very common in the French countryside. Its fructification organs (apothecia) draw on the thallus small cracked lines, which can be simple, branching, or even in the form of stars (see Ill. 4) and perfectly resemble script or punctuation (poet Brenda Hillman compares the edges of *Flavopunctelia soredica* to tildes ~ and *Ramalina menziesii* to number signs #), symbols, and even chromosomes.

They are called lirelles, from the Latin *lira,* the furrows made by a plow, as if they were drawn by a plowshare; here they return us to the imagination (*lira* is the root of *délire* in French, "delirium," that is, leaving the furrow) and to writing (in Latin, the furrow of the plow is also the *versus*, the "verse" of poetry). Carpeting branches or wrapping around the twigs of deciduous trees, these lichens are an invitation for deciphering: read the lirelles, see them and touch them, like braille – the title of a collection by German poet Hans Magnus Enzensberger (born in 1929):

Figure 8 A pattern of lirelles of a lichen of the *Graphis* genus. © Vincent Zonca, Rio de Janeiro, 2020.

The lichen has its graphics
its inscriptions its encoded
writing that describes
a prolix silence:
Graphis scripta.[21]

"It carpets with indecipherable writing the supports that it chooses: lower or upper case; inlaid or in relief, linear, forked, Chinese, cuneiform," writes poet (and lichenologist) Camillo Sbarbaro in his turn. In 1789, for the great naturalist Jean-Baptiste de Lamarck, "the surface [...] offers a great number of interrupted crevices, which make it seem strewn with small blackish lines, arranged in various directions [...] somehow resembling Hebrew letters or Chinese characters."[22] Or again, for Henri Michaux (1899–1984), there is the hallucinated calligraphy of his sketches. Another Belgian painter/poet, Christian Dotremont (1922–1979) linked lichens to his graphic research. Discovering the Finnish Lapland, he was overwhelmed by the minimalism of the landscape, as well as by the immensity of the snow that made him think of a gigantic white page. Here and there on the arctic snow, lichens survive (*Cladonia stellaris*: reindeer lichen) as well as a few sparse plants, like abstract ink marks on a white page.

As Michel Serres notes, the word "page" comes from the Latin *pagus*, which means "the boundary stone of a field, embedded, planted into the ground, marking its border, a half-buried stone, limit, stela that in its first version stood there on the tomb of the recumbent ancestor."[23] The page is thus a territory, it is the peasant's landscape. Because, for Dotrement, creation is landscape, and vice versa. It is "a living thing" and must reclaim the instinctive part of the human. Hence this primitivist tendency, this blurring of boundaries between nature and culture. In the 1960s and 1970s, these visions gave rise to his "logograms," a writing system done with Indian ink that gives rhythm and gesture to the traditional Latin alphabet, a kind of choreography, exaggerating our shared tendency toward illegible handwriting and renewing the practice of the calligram and of Asian calligraphy.

Some logograms have titles like "Rhythms of lianas, writings, lichens" or "Yesterday again, I said to Baudelaire: lichen, how beautiful!" Over the course of his travels, he even traces "logosnows" and "logoices" on the arctic ground.

Representations

> "The most common lichens, the flat lichens, are a bit like palette knife paintings dispersed throughout the landscape. Nature, what distant (and celestial) painter had so carelessly cleaned his spatula there?"
> Pierre Gascar, *Le Présage*

A matter of multiple colors and textures, a form of landscape, a form of world, lichen has inspired numerous visual artists. But it may be less pictorial than graphic, plastic, ceramic. Crustose lichens make magnificent mosaics.

Relatively few painters, in fact, have represented lichen (there are Albrecht Altdorfer, one of the first landscape painters in the West, Arcimboldo, Escher, Pitxot, Saby, Hercule Seghers, and a few Romantic watercolorists, as well as the painting tradition of China and Japan). Many more artists have represented it using texture and line: drawings and engravings (Gérard Titus-Carmel, Thomas Fouque, the scientific plates of Ernst Haeckel, the lithographs – drawn on stone, like lichens – of Yves Chaudouët); ceramic sculptures (Leo Battistelli, Cathy Burke), in wax (Simone Landwehr-Traxler) or paper (Amy Genser, Laura Miranda), weavings and embroideries (Claudia Losi, Laura C. Carlson, Amanda Cobbett, Hannah Streefkerk).

The German word for lichen, *flechte*, comes from the Greek πλέκειν [*plékein*], "to braid," "to interlace." One species, *Ramalina menziessi*, can almost be mistaken for lace. From greenish yellow to greenish gray in color, it can grow to one meter in length and hangs from the bark of trees. Well known in North America, it is called "lace lichen" and resembles fine, frayed lace or a spider web in tatters (it was named the California State Lichen on July 15, 2015 by California governor Jerry Brown as a state symbol, like the California State Animal, the grizzly bear). Erasmus Darwin, Camillo Sbarbaro, and Pablo Neruda describe lichen as "tapestries," while Balzac describes it as a piece of silk:

> This wild quay was covered with many species of lichen, a beautiful fabric shimmering in the dampness, which appeared as a magnificent curtain of silk.[24]

*

Visual artists today pay very special attention to the natural world and to lichen. Returning to traditional craft techniques – embroidery, enameling, engraving – they can delicately express the complexity and fineness of the thallus.

Italian artist Claudia Losi (born in 1971) captures its fine texture and delicate feel. Since 1995, in her work-in-progress entitled *Tavole vegetali*, she has been making pieces of embroidery (already more than thirty of them) on fabric and sugar, representing the fantastic, colorful forms of crustose lichens that she has photographed over the course of her travels. The series constitutes a private journal, an inventory, a microcosm:

> These plants [...] can represent the dynamics of growth, expansion, and death that, on a larger scale, are the same as the ones that regulate the natural landscape, as well as human landscapes and relationships. There are metonymies of complexity, there are "spots of growth" [*macchie de crescita*]. Each finished work ideally becomes a page in a kind of herbarium that recalls collections from the nineteenth century.[25]

The composition of this herbarium, these works, is as slow as the growth of lichen. It is the long work of embroidery, and this artisanal technique allows for the expression in minute detail of all the contours and effects of the material (I have found the image of embroidery again in the writings of John Ruskin and lichenologist Guy Nearing, and this link with artisanal objects in botany manuals: certain parts of lichen seem "pierced with a needle," "made like a thimble," or like a "bezel on a ring"). Lichen reconnects with the skin – recall the lichen of dermatological diseases – through a second kind of skin, the inoffensive envelope of clothing, in delicate, exquisite pieces of embroidery forming abstract micro-landscapes as so many brocades or designs. In 2018–2019, the US artist Laura C. Carlson also created a series of delicate embroideries entitled *We Are All Lichens*.

In his series *Diffusion* (2016), the French artist Mathias Tujague (born in 1980) offers a series of hybrid sculptures both mineral and vegetable. Grafted to various structures made of crystals of borax, or sodium borate, a mineral present in dry lake beds and deserts (especially in Death Valley, California), lichen finds an analogue for expressing resistance and permanence. The various colors of these crystals (green, orange, yellow) play with and mimic the polychromatic lichens. Lichen and its crustose rediscover in these encrusted sculptures a sharp, crisp form. All of this results in an irrepressible desire to touch the dry lichen and make it crunch between the teeth, under the foot. Responding to all the shades of colors is the opposition between mineral and vegetable, between death and life, which escapes, by a thread, through means of the crystal, like a form of coral (a material Mathias Tujague is also fond of). Lichen achieves the dignity of sculpted stone here (silversmiths commonly use borax for scrubbing metals in order to isolate precious stones). These "laboratory rovings" are a meditation on material and object.

Just as for Claudia Losi, nature and the sciences are fundamental sources of inspiration for Tujague's artistic practice. Natural materials such as plane tree leaves, crystals, pebbles, moss and lichen, charred wood, dead coral, and terra-cotta are reproduced and transformed – hybridizations or scenographies, alterations in materials or scale – in sculptures and installations that are formal and poetic experiments inviting reflection on their cultural uses. The magnificent *Turricules* (2015) series thus presents *"verres de terre"* ["earth glasses/worms"]: worm excrement was collected, then partially enameled through baking, creating astonishing branching sculptures, the sleek black of the enamel letting the granular brown earth show through.

Embroidered, enameled, mounted on crystal, but also drypoint engraved: lichen inspires Thomas Fouque (born in 1986), who imaginatively recreates botanical plates, bestiaries, herbaria, and cabinets of curiosities. Recalling the heights of botanical engraving in the nineteenth century, his very small format works (8 x 8 cm) try to rediscover the subtle metamorphoses taking place in the infinitely small world of our gardens by presenting the mystery of life cycles and forms. As "ephemeral islands" suspended on the white page, his lichens thus stand out like writing, like tiny granulations that depict worlds/forms. As Bachelard, cited by the artist, wrote:

> The botanist [...] ingenuously uses words that correspond to things of ordinary size to describe the intimacy of flowers. [...] He is a fresh eye before a new object. The botanist's magnifying glass is youth recaptured. It gives him back the enlarging gaze of a child. With this glass in his hand, he returns to the garden. [...][26]

This magnifying glass/gaze of the child and botanist, which turns these miniatures into worlds, is also that of the Swedish artist Oscar Furbacken (born in 1980). In 2009, he chose to work close-up by enlarging into an immense photographic print (5 x 5.60 m) a lichen he found on a tree close to the Stockholm metro (the famous *Xanthoria parietina*; see Ill. 11).[27] The detail opens into a dream, the change of scale letting a "surreality" appear:

> It is a question of scale, of status, of disgust and beauty. Instead of a subtle, rarely noticed natural element, the spectator is

> confronted here with lichen expanded to absurdity. Each crack and wrinkle reveals something. The coral formations and colors of the lichen take the shape of a psychedelic sun, a sulfur lake, or perhaps a topographical map of an interior landscape.[28]

The artist's fascination for the infinitely small – mosses and lichens are present in almost all of his works, sculptures, photographs, videos, or installations – conveys a spiritual dimension: the desire to show the invisible part of creation, to invite romantic contemplation of nature's transcendence and empty landscapes (*Wastelands*, 2014) by using new techniques. In 2008, in *Les Petites Merveilles de Paris*, as a kind of trompe l'oeil, he covered the white walls of the Pavé d'Orsay exhibition space with touched up and enlarged photographs of lichens and mosses found in and around Paris. In 2012, lichens and mosses sprang up on the interior of the Katarina Lutheran Church of Stockholm, playing with the baroque elements (see below, Figure 11b) of the architecture and reenacting the third day of Genesis (*Ascension* or *Beyond the Second Day*).

> Should we mention the lichenous efflorescences found here and there on certain neglected works of art, in particular on the frames and sometimes even the canvases of the large paintings [...] that decorate Jesuit churches? These lichens only ever cover quite small surfaces and moreover, do not attack the material to which they adhere. They affix on the paintings a seal of antiquity, of authenticity, and, in short, bestow on the religious art nature's quiet respects.
>
> Pierre Gascar, on the subject of Venice[29]

In 2012, Oscar Furbacken's close-up work was followed by a series of videos entitled *Close Studies I–VII*. An explorer of the familiar, as Rousseau had expressed it in another age, he exploits the possibilities of video and its latest innovations in order to capture the infra-movements of lives tucked away in the interstices of our cities (see below, Figure 11c):

I play the role of a maker of nature films exploring distant countries. But instead of traveling in remote regions, I chose to focus on lichens and mosses in close proximity to my daily life. In this project, I begin with the principle that if God exists, we should be able to find traces in the small living beings that surround us. Is it possible for video recordings of a given place to retain its spiritual energy? For the camera to extend beyond my physical field of vision, I developed a jib for extremely close up micro-movements.[30] A major part of the project consisted of designing, constructing, testing, and adjusting this equipment, as well as finding adequate accessories. Once in place, I had fun finding a presence thanks to the point of view and movement of the camera.[31]

Thus enlarged, lichens and plants become a world, are transformed into hyperbole – forests, cliffs, escarpments, lunar surfaces – and are revealed in very slow motion as so many silent aerial views. Also on the subject of lichens, Bachelard wrote: "Miniature is one of the refuges of greatness. [...] Miniature is an exercise that has metaphysical freshness; it allows us to be world conscious at slight risk."[32] In the Far East, the miniature reinforces the intensity of a subject's qualities. Botanists have compared one variety of lichen, *Rhizocarpon geographicum*, to a map of the world: at the surface of its "crazed, leprous crust" [...] "interstices form black anastomosic lines in various directions, the main ones somehow resembling rivers, thus giving [...] the lichen [...] the appearance of a small geographical map."[33] Jean Giono wrote:

> It looks like a big country. You see those green patches encircled in black, there, the reddish plains with the little brown line separating the fields. The seas, rivers, oceans with their colors and forms.[34]

What is small becomes big, what is close becomes distant. In 2013, on the roof of a Nolhaga Park building, there were sculptures that reproduced enormous lichens (*Höjdarna* – "The Heighteners").

When I met Yves Chaudouët (born in 1959) in Paris on April 4, 2018, he greeted me by saying that he would soon be leaving his studio in the 18th arrondissement for the provinces and the countryside. Bound for: fresh paint, folded works. The big city is not, in fact, the favorite place of this artist, who has always cultivated a taste for the country, a nomadic existence, the movement of rivers. He is also errant in his artistic practice. Although it is very coherent, it is not afraid of wandering and mixing very different disciplines: music, drawing, the novel, the theater, film, still and moving images, the sciences. Yves Chaudouët is the maker of an open, polymorphous body of work, and is happy to enlist artisans and specialists in order to enlarge the spectrum. His flight from urban centers and taste for the sciences are explained by a formative event. In the 1990s, he decided for the first time to leave the city. It was in this context of "survival" that he made what he calls "the landscape experiment":

> One day in 1995, I went out. I wanted to go to see in the actual landscape that my vast romantic fantasies were based on. Equipped with an optical camera as serious as my intentions at the time, I squatted down to touch lichens, peony sprouts, mushrooms, ferns, springs, as well as all sorts of mundane curiosities that grow between the paving stones in cities and proliferate in the wild around Cantal where I lived then. Guided by some of my rural neighbors who, like wise grammarians, taught me two or three essential things for reading the forests, valleys, waterways, and hills, I penetrated their secrets, trading colors and lenses for more botanical tools. I never returned from this landscape "experiment."[35]

Picking up on this theme, Chaudouët developed a "handwriting" of the landscape, rendered through a series of photographs that captures what appears along roadsides.

A Challenge to Representation

This interest in nature and small things is part of a wider curiosity, even affection, for what is laughable, disdained, neglected: roadside plants, deepwater fish, languages, forgotten minor authors. It was John Cage especially who initiated him into the mystery and the aleatory nature of nature by way of mushrooms. Thus, in the heart of Cantal in 1995, Yves Chaudouët found himself face to face with the mystery of plants and did a first experiment with lichens. What struck him was the mysterious growth of these organisms, the challenge of representing their complicated textures. It was also an affinity for a tiny, marginal presence that resists.

He then created a series of small format photographs (6 x 6 cm) on this theme, with titles that evoke the landscape: *Mountain*, *Crest*, and so on. As always, the technique used was thoughtfully chosen and involved extreme representational challenges. In this case, Chaudouët used Cibachrome (Ilfochrome), which required the help of Roland Dufau's photography studio, the last of its kind in Paris equipped for this process. Unlike other diapositives (Kodak-type), Cibachrome allows the lichens to be presented without a filter, with an almost clinical neutrality and coldness, rejecting all pictorialism. "The print withdraws, disappears behind the represented subject," Chaudouët told me. Moreover, Cibachrome paper is unusual in that it already contains the colors, in different layers, that the light then comes to burn. In this sense, it is a matter of true polychromatic sculptures, hollowed out between the veneer and the support (between the superior cortex and the substrate of lichen, as it were), which allow for the ideal representation of lichen's contours.

In 1998, Chaudouët experimented with lithographic techniques. This time he worked with Christian Bramsen's Atelier Clot (lithographic experts since the nineteenth century and Degas' time) for the printing of a portfolio entitled *Lichens* (see Ill. 13).[36] Once again the traditional – even outdated – technique responded to the challenge of the landscape and the modernity of the view. The seven plates show seven lichens, enlarged, and finally, abstract, dreamlike. Are we seeing a single thallus or several entangled thalli? What are these lichens, what types of lichens? In fact, it is difficult to *recognize* what we see, except as an abstract game with colors and textures. Thanks to the lithograph, the ink collects in certain areas as it dries, creating webbing, wrinkling effects, "elephant skin!" exclaimed the artist, thinking of the furrows of the pachyderms that allow them to store water (the *Diploschistes scruposus* lichen with its thick, light gray thallus almost perfectly resembles elephant skin). The thickness and distribution of the ink here replaces the luminous Cibachrome sculptures in working the texture. Sometimes the lithograph lets the white of the paper support show through: the transparency of lichen. This lithographic gesture is not innocent: as with lichen, the color is deposited on the stone, then it is printed onto the paper; furthermore, like lichen, which "licks" its support, the lithograph involves less work, less of a struggle against the metal, as with engraving, and more actual "sensuality" (that term is the artist's). It involves gently depositing the ink, with a wash-paint brush (thus, to wash, to bathe), onto the limestone, caressing it. From this pursuit of texture emerges the dream: colors and forms are beautiful infidels in the drawing of these true-false lichens. The motley colors are vivid, bordering on impressionism (the impressionists were equally passionate about lithography). There are often three colors, complementary or creating "thermal shocks" between their cool and warm tones: lichen kitsch.

"Dream Flora"
Camillo Sbarbaro invited me to see lichens through the lens of existential quest, within a psychological context (projection, prenatal fantasy, and so on): ecology as rescue, as hope. The Italian poet's view is radically subjective; he sees them to better see himself in them. As Pierre Gascar writes:

> What lichens represented, these plants with their vague forms, sometimes creeping or flat as stamps, sometimes hanging in streamers, their softness to the touch: didn't they border on

repulsive? The creations of the unconscious, dark thoughts, fantasies. [...] Are they truly there or do we see them only in dreams? We come to think of them as "sight's vegetation."[37]

The "image-making forms" of lichens are so many mirrors in which to see ourselves; we see in them forms that reveal us, that reveal our unconscious. As an organism that stimulates the imagination of writers – let us recall the magnificent descriptive images of them by Gascar, Thoreau, or Caillois – and scientists, it is not surprising to find lichen evoked by the surrealists beginning in the 1950s to better disclose the secret margins of our "ego," the vitality of our interior world. Ruderal: the unconscious.

*

The artist Bernard Saby (1925–1975), who was a singular and major personality in the surrealist art world of Paris in the 1950s and 1960s, has almost vanished from memory today. It was thanks to the writers Michel Butor and Armand Gatti that I was able to track him down indirectly in a library (the Bibliothèque Nationale de France lists only two references to his work). Following the Second World War, he developed an original language, rejecting surrealist representation in favor of abstraction. His paintings and engravings can be recognized by the dense and narrow lines that run through them and draw complex, mysterious architectures, reminiscent of abstract expressionism, of André Masson, but also of the thalli of lichens that intermingle and superimpose themselves.

Passionate about zoology and petrography as a child, Bernard Saby began his studies at the College of Sciences in Paris in 1945 and specialized in botany. The following year found him volunteering at the National Museum of Natural History in Paris, working with two specialists in algae (Lamy and P. Bourrelly). He was interested in lichens in the usnea family. He classified files and collections, and made enthusiastic botanical expeditions around Paris, which earned him a publication in the *Revue bryologique et lichénologique* when he was just twenty-one years old.[38] At the same time, he was pursuing advanced studies in serial music and composition with René Leibowitz and Pierre Boulez, born the same year, and had fallen in love with China and Taoism. Eventually, in 1947, he decided to devote himself to art. Abandoning music, he took up another of his childhood passions: painting. His various interests nevertheless fed his pictorial language throughout his career, punctuated with times of depression. "He was also a musician and had been part of the first twelve-tone group after the war, along with Boulez, Stockhausen, and others. He was a sinologist and lichenologist and with all of that, completely at sea in daily life," wrote Michel Butor.[39] He was interested in Poussin and Seurat, and in the late 1940s, kept company with Armand Gatti, Henri Michaux, John Cage, then Zao Wou-Ki in 1953. It was his 1963 exhibition at the Galerie de l'Oeil that brought him his first real public success. Notably, he impressed Butor, whom he encountered again on this occasion. Butor became a faithful friend and acknowledged him in various texts, in which he called Saby's canvases "lichens."

The organic image is sometimes explicitly evoked by Saby himself, and linked to a form of hope: "Here, a vegetal hope, coming not from itself but from centuries past – fragile and ephemeral like a single day – was sketched. [...] At the surface, a few kelp will indicate the spot of an anticipated shipwreck" (in the margins of the first charcoal sketch). Abstract images like architectures (Butor speaks of "construction games" made by the "trowel of the mystical mason") make one think of lines of force or organic proliferation. Open to depths, to perspective, they are structured from micro-cells that resemble lichens.

They open onto a completely surreal form of depths. The painting is an exercise of the the unconscious, a "knowledge through gaps." "One day, in my paintings," wrote Saby, "a depth started to appear, and I plunged into it, I threw myself headlong

A Challenge to Representation

Figure 9 Bernard Saby, *Untitled*, 1958, oil on canvas, originally in color, 130 x 97 cm © Galerie Les Yeux Fertiles, Paris.

into its exploration." From there, as well, followed "artificial paradises" (mescaline or hashish), but also Taoist spirituality, to help make these primal, interior visions emerge.

> You were the painter of the desert, and its monsters that have never ceased to be, and its forms always on the point of disappearing into the ones that preceded them and whose original ghosts are woven over all your canvases.
>
> Armand Gatti[40]

*

For the surrealist writers, natural elements constituted a means for releasing the imagination, a tool for freeing the unconscious. Roger Caillois cultivated "natural images," Michel Butor saw in lichenology "a methodical exercise of the imagination."[41] The whole forest hums with voices from the unconscious:

> On my first shelf the alphabet of lichens, on the second the syllabary of runes, on the third the dictionary of shoots, on the fourth the encyclopedia of looks, and on the top shelf the anthology of breaths.
>
> From my maintops chirp the topmen, from my ropes tumble the mosses/cabin boys [*mousses* means both], from my bowsprit flap the flags of snow.[42]

Butor was passionate about botany.[43] He first encountered Bernard Saby early in his artistic career, no later than 1961, and let himself be initiated into the marvels of the world of lichens.[44] They shared a fascination for dreams and for Taoist spirituality, and it was this organism that Butor evoked most often when he spoke of his friend. What he captured was the "magic" of its power to make images and to stimulate the imagination:

> *Lichen, for me, is natural painting.* It is painting that is done all by itself. I am fascinated by lichens. When I look at walls, I find them extraordinarily beautiful. I am very familiar with them around my home in Nice. And from time to time I go to look at them where they grow. Sometimes there is an old wall where a lichen has grown another patch or a promontory that has added something magnificent. Also when I am to speak on Japan, on Japanese art, I will describe in as precise detail as possible the Ryōan-ji garden, that garden with its fifteen rocks that occupy the sand, with a bit of moss. And on the rocks, there are patches, lichens. I adore lichens. *They play a fundamental role within all that magic.*[45]

If, for Rousseau (Butor also, interestingly, lived near Geneva), lichen opens into reverie (see below, p. 97), here it is a surrealist method, a magical practice. And in leafing through Butor's novels and poems, which are just one and the same way of conceiving of writing in a visual and transitory space, although lichens are not found everywhere, they systematically prompt a vision, an opening toward depths, toward the invisible:

Of course, it was only a very imperfect image; the slate roof shingles seen from above, the smoky chimneys, the dark bricks, the macadam of the roadways seemed to form a large gray lichen with reddish, wrinkled, rough streaks, covered with foam, like one finds on the rocks along the coast of cold seas.[46]

Are you going to let it implant itself and grow on this face as well, this lichen of suspicion that made you hate the other [...]?[47]

I descend line by line the steps of the large water tower where lichens continue their living paintings.[48]

The lichens on the rocks took up their paintings again: flowers or nebulae, crowds or maelstroms.[49]

Aerial views, suspicious signs, "living paintings": lichens transfigure reality.

*

Among other artists of this period, we find the same interest. Dutch lithographer Maurits Cornelis Escher (1898–1972) was fascinated by lichen; in the background of his famous 1961 *Waterfall*, he represents giant cladonia that add to the strangeness of the scene. Antoni Pitxot (1934–2015), Catalan painter and friend of Dalí (Pitxot was born in Figueras thirty years after Dalí), developed a pictorial language marked by surrealism and the establishment of his studio in Cadaqués in 1964.

Fascinated by the irregular rocks covered with lichen that border the Catalan coast, he used stones he found to create sculptures, before reproducing them on canvas, all in dialogue with various myths and references, like Shakespeare's *The Tempest*. Cadaqués stones lend themselves to pareidolia; they prompt anamorphic, anthropomorphic, and even allegorical visions. Halfway between Arcimboldo and De Chirico, certain paintings, with classical compositions and themes (nudes, portraits, self-portraits), represent mineralized or even lichenized bodies and erase facial features. These astonishing visions also recall the grotesque monsters and worked materials of the Italian mannerists. In his 1974 *Self-Portrait*, the rocks covered with a yellow and brown lichen compose a stern, melancholic face, like an ephemeral monument. Moreover this face is detached from a background that suggests a sky but is also only a mineral and lichen surface. Such mineralized bodies evoke lichen as cutaneous manifestation, dermatological lichen (we find this again in his 1975 *Female Nude*, his *Torso against a Yellow Background* and his *Three Graces* in 1997). They play with classical references, muses, and myths that have become grotesque. His landscape paintings also give prominence to lichens, as in *Cypress and Lichen* (1972) or *Grass and Cellist* – although they do not have the degree of abstraction and hallucination found in the background landscapes of paintings by Dalí or Escher – just as do some of his sculptures and installations. But his most impressive work is no doubt the full-length portrait of the patron saint of Catalonia and the Royal Academy of Beaux-Arts, *Saint George* (1976), a life-size representation (almost 2 x 1 m), whose noble posture and clothes are found to be colonized by a bright orange lichen covering both skin and tunic (see Ill. 5).

Music = Mushroom

Lichen's power over the imagination, combined with surrealist conceptions of creation and Taoist thought, also showed up in the United States in the same period, where new models of musical composition especially were developing.

Saby, and this relationship to Taoism, musical composition, and botanical studies, immediately bring to mind John Cage (1912–1992) and Black Mountain College – the home of unprecedented artistic experimentations in the United States beginning in the early 1950s. There, in the summer of 1952, John Cage created his first untitled "happening," accompanied by David Tudor on piano, the poems of Charles Olson, and the monochromatic white paintings of Robert Rauschenberg (1925–2008, born the same year as Saby). The audience was located at the center of the action.

This happening and the white monochromes were connected to his most famous work, created in August of that same year: *4'33"*. In this piece, with its empty white score, the music is created from the silence and as though in the negative, from the accidents of the environment each time the piece is performed: silences from the instruments but noises from the room, the audience, the outside.

In the late 1940s, Cage had a crucial experience when visiting a soundproof chamber at Harvard University. In the silence of this echoless room, the music of his own body became evident to him; this was his revelation of the impossibility of absolute silence:

> From Rhode Island I went on to Cambridge, and in the anechoic chamber[50] at Harvard University I heard that silence was not the absence of sound but was the unintended operation of my nervous system and the circulation of my blood. It was this experience and the white paintings of Rauschenberg that led me to compose *4'33"*.[51]

Such a meditation on silence is also related to the Far Eastern thinking that was very much in vogue at Black Mountain College at that time. Beginning in 1951 moreover, Cage took courses with Daisetz Teitaro Suzuki at Columbia University in New York. Suzuki was a Japanese teacher and translator who played a major role in bringing Zen Buddhism to the West after the Second World War.[52]

Cage's long walks in forests in the 1950s were also part of this. The sound of the slightest noises in nature, solitary meditation [...] and botanical research into small things – mushrooms and, failing that, lichens – at once recall Thoreau, whom he adored, and Zen.[53]

In fact it was precisely at this moment – summer 1954 – that Cage abandoned New York City for the country, Stony Point, New York. This retreat was not solitary; he was accompanied there by friends, among them the botanist and writer Guy Nearing (1890–1986). Cage developed an environmental awareness – he worried about nuclear waste in the context of the Cold War.[54] And he then devoted himself with a passion to collecting mushrooms and lichens. Fungi, lichenized or not, store radioactive elements; they also provide the name of the form of atomic explosions (mushroom clouds). Cage was fascinated by their arbitrary distribution, by their subtle presence, which involves listening carefully and being quiet; they would inspire his musical and poetic compositions that cultivated chance and the aleatory.

> Ideas are to be found in the same way that you find wild mushrooms in the forest, by just looking. [...] I have come to the conclusion that much can be learned about music by devoting oneself to the mushroom.[55]

Thus Cage tells how he fell in love with this organism by flipping through a dictionary, because the word *mushroom* is the one that immediately precedes *music*: "their link is arbitrary; the two words are neighbors in many dictionaries" (*For the Birds*, 1981). Just like Sbarbaro with lichens, Cage collected mushrooms (his impressive collection is now found at the University of California at Santa Cruz). In 1959, he even appeared on an Italian television quiz show as a mushroom specialist, identifying mushrooms; he won first prize (thanks to which he bought a new piano!). Three years later he founded the New York Mycological Society with Guy Nearing and two of his friends, author Lois Long and mycologist Alexander Hanchett Smith.

> It was in the fifties that I left the city and went to the country. There I found Guy Nearing, who guided me in my study of mushrooms and other wild edible plants. With three other friends we founded the New York Mycological Society. Nearing helped us also with the lichen about which he had written and printed a book. When the weather was dry and the mushrooms weren't growing we spent our time with the lichen.[56]

Here is an account of the same scene from another perspective:

Music = Mushroom

The first time that I went mushroom hunting with Cage, it was too dry to find them. Instead, Cage began picking up rocks, pulled out a magnifying glass, and started to look at lichen formations. From a distance, these were hardly more than spots of color. However, through the magnifying glass, each piece revealed the complexity of its design, as specific as the brushwork on a canvas by Jasper Johns. Cage knew how to transform any walk in the woods into a voyage of discovery, whether or not mushrooms were part of that discovery.[57]

In 1972, with his two friends, Cage published *The Mushroom Book*, a very unusual book printed on Japanese paper, combining mycological studies with visual art and poetry. Some texts are lithographs and appear as concrete, visual poems; expanding from the bottom to the top of the page, they grow in a haphazard way, as though advancing in patches, in puddles. In 1983, it was followed by a *Mud Book*.

The meditative confrontation with silence while mushroom gathering is a kind of potential, incongruous extension of the *4'33"* piece. Moreover, lichen evokes not only the painter Jasper Johns but also Robert Rauschenberg, whom Cage greatly admired, with whom he collaborated. In 1953 Rauschenberg made his *Elemental Paintings* series that included *Dirt Painting (for John Cage)*, a box filled with dirt combined with lichen and soluble glass. The painting was left exposed to impurities: mold, lichens, moisture. The pictorial gesture was relinquished to better reveal the very fabric of the canvas and the natural accidents that could affect it. In the words of the artist, collaboration with the material took precedence over the conscious and premeditated gesture.

In the Black Mountain College era, accidental and collaged vegetal motifs observed in nature thus appeared as images in a musical, pictorial, and poetic language, as silent, aleatory principles of composition: botany as spiritual exercise, at the intersection of romantic (transcendental) experiences and Zen, and at the heart of ecological thinking.

*

In the decades that followed, musical composition sought to practice serial discontinuity by making use of the sciences and the earliest computers.

In the late 1960s, György Ligeti (1923–2006) developed a technique inspired by the phenomena of biology, molecular physics, and mathematical laws, creating textural sound effects that evolved according to complex, rigorous principles of transformation, reminiscent of lichen proliferation and the mathematical model of "cellular automata."[58] In the 1970s and 1980s, the French and Greek composer Iannis Xenakis (1922–2001) developed a type of polyphony and writing based on a principle of "arborescence" (that he linked with "bushes"), structural proliferation in which the natural model was acknowledged, even while remaining very abstract. The sounds of nature had influenced him since childhood.

> And I infinitely loved nature. I would go by bicycle to Marathon. At the place where the battle supposedly took place, there was a burial mound with a bas-relief of Aristocles, and there I rested for a long time soaking up the sounds of nature, cicadas, the sea.[59]

As evidence, many of his works have evocative titles: *Jonchaies* in 1977, *Lichens* in November 1983, both for orchestras. In this latter piece, performed by ninety-six instruments, the percussions are omnipresent and explode into an intense, exuberant music, like a gushing spring, playing with glissandi and testifying to an energetic, Dionysian vision of music when it expresses the life force.[60] The technique of "sound halos" produces foregrounded melodic lines (ascending, descending) that create reverberating effects. The music that Xanikis developed in this period was inspired by the aleatory movements of molecules (Brownian movements) and used mathematical calculations of probability to develop a new type of serialism.

The serial music of the years 1950–1980, exploring the limits of sound and aleatory logic, thus found in the sciences (mathematics, physics, biology) a favorite means of expression. For Saby, Cage, Ligeti, and Xenakis, the fungus/lichen seems one of the models for this experimental era of music.

The Far East, Mosses, and *Wabi-Sabi*

> "While Japan appreciates them, nurtures them, and cultivates them, the West usually ignores or eradicates them. Sworn enemies of the lawn gods, [mosses] are hardly the delight of more than a handful botanists."
> Véronique Brindeau, *Louange des mousses*, 2012

Lichens, like mosses, enjoy great success in Japanese culture. Far from being considered weeds or a disease, they are treated with care and take part in the codified design of the Zen garden – and the haiku poem – in the same way as rocks, water, sand, gravel, carp, lanterns, dirt or stone paths, tea pavilions, lotuses, pines, and cherry trees do. Their presence there is directly tied to that of mosses: the Japanese equivalent of the word "lichen" remains a technical, recent term (*chui*); the word that designates moss is the one generally used (*koke*). Many species of lichens thus include the word "moss" in their names.

*

While walking in the Marais district in Paris the other day, I discovered, winding through those mineral streets, some lichen decorating a shop window – a Japanese fashion boutique, *Pas de Calais* – delicately featured between wood and stones.

*

In Japan, lichen and moss are truly part of the garden: two thousand five hundred species of moss grow there (out of the twelve thousand known species worldwide). The humid climate, tied to the monsoons of East Asia, is very favorable to them. With their naturally moist, mucous textures, they embody the porosity among the elements so dear to the Japanese aesthetic moreover. As Véronique Brindeau describes in her *Louange des mousses* [In Praise of Mosses], there may be as many words in the Japanese language for naming clouds as for naming mosses: "the smallest pocket guide for moss lovers includes more than three hundred of these common names, all full of lanterns, brushes, squirrels, and frost." This lexicon tends to make familiar, to popularize these "grasses of memory" as they were called in ancient Japanese texts; they are objects of fascination and contemplation (of "full consciousness"). In 2011, the book *Mosses, My Dear Friends* by Hisako Fujii sold more than forty thousand copies. Let us also think of the practice of *kokedama*, those "spheres of moss" that hold a single growing plant, as well as the Japanese garden, which can bring together more than one hundred different species of mosses. Kōinzan Saihō-ji, the Zen Buddhist temple in Kyoto nicknamed Koke-dera, "Moss Temple," constitutes a fabulous achievement in this respect, restored by Musō Soseki, the famous fourteenth-century master of landscape gardening.

That is because moss and lichen can be linked to a secular principle, both ethical and aesthetic, called *wabi-sabi* (侘寂). Structuring the Far Eastern perception of the world, it appeared as early as the twelfth century, beginning with Taoism, and then the Zen Buddhist philosophy that flourished in feudal Japan in that period. This principle combines two concepts, hard to translate into another language even if they now enjoy worldwide success in self-help and spiritual development guides: *wabi*, which means something like the spiritual quest for simplicity and solitude, asceticism and withdrawal; and *sabi* (almost like Saby, the painter), meaning something like cultivating imperfection and impermanence, the marks of time, an aesthetic of wear and tear, an object's patina, alteration, tarnish, a deterioration that is nevertheless "not decrepitude or blemish but a pact with the passing time."[61] It is a complete reversal of Western values. We are very far here from the quest for permanence and aesthetic perfection, as well as the perception of *tempus fugit*: the passing time is not seen or experienced as fleeting, lost, or cause for nostalgia, but as constructive advance, the normal course of things. Without resignation, it is a matter of accepting and welcoming the world as it is, in all its temporality, to better be included within it. These two concepts are related: the first allows for discovering and experiencing the second. This ancient principle of *wabi-sabi*, which thus advocates a return to peaceful imperfection and simplicity in order to better reintegrate into the natural cycle, involves both a lifestyle (an ethics) and a model for beauty (an aesthetics) fundamental to Japanese culture and still present today.

According to André Leroi-Gourhan:

> That is what leads aesthetic tastes toward things that remain close to their material; that is why a tea bowl must make one think of a natural object. A garden stone covered with moss only needs humans to value it; beyond that, it has its own life, its own rhythm. A pothook blackened by smoke, a wooden wall eroded by water, these are so many pieces of evidence marked by the passage of time. Each is a portion of the life cycle and on each, as on the rusted iron kettle, what provides that aspect of duration, of regulated flow, what provides that patina, is *sabi*: rust. "Patina" is not an exact translation of the term; "rust," extended to wood no less than iron, is closer to the etymological sense of it. The moss on the garden stone is "rust," the crust of soot on the pothook is also "rust," as are the burnish of the wood in the gallery, the lichens on the plum tree, and the mold on the fence.[62]

In this sense, lichen is *wabi*, a modest, slow, laughable organism: a detail, minutia. It is also *sabi*: "rust" for Leroi-Gourhan, stamp of life, experience of the body ("rust," which means conversely "inaction," "bad habits" in Latin; there are images of oxidation in Jaime Siles' *Late Hymns*, see below, p. 137 – moreover, lichen *oxidizes* rocks).

Here lichen is once more associated with temporality. However, there is no resistance or opposition here, ideas foreign to the Japanese principles.

> The tree's answer to the wind's force echoes the *Taoism of lichens*: don't fight back, don't resist, bend and roll, let your adversary exhaust herself against your yielding.
> David George Haskell[63]

Above all, it is lichen's longevity and impermanence that are praiseworthy (age, venerability, antiquity as values) and, especially, its ability to *show* the passage of time. It shows less resistance than energy, than assent to the world, intensity, eternity – and a taste for change. With its very slow growth, lichen articulates and measures the flow of time in nature, and is both a sign of eternity and precariousness, as well as the patina or rust that marks things.

In classical Chinese poetry, the traces of footsteps on fresh moss (or moss that grows on the traces of footsteps) is a recurring motif: passing layered over passing. It also marks the slowness of elegiac time after the departure of a beloved:

> At our gate, where you lingered long,
> moss buried your tracks one by one,
> deep green moss I can't sweep away.[64]

Or it can be a positive motif, celebrating Japan's eternal prosperity in the national anthem:

> May your reign
> Continue for a thousand, eight thousand generations,

Until the tiny pebbles
Grow into massive boulders
Lush with moss [*koke*].⁶⁵

Supports colonized by vegetation – stones, branches – thus emerge imperfect, lose their cohesion, their luster, their neatness. Their surfaces take on many colors and are concealed, their brightness diminished. This Japanese fascination for the alteration of the world was brilliantly described and analyzed by Jun'ichirō Tanizaki in his *In Praise of Shadows* (1933). Lichen creates this disturbance in things, conducive to mystical meditation, that Tanizaki also sees in Chinese jade, in the shimmering of candelabras on the walls and lacquered furniture in houses, as well as in a dark bowl of miso soup.

> As a general matter we find it hard to be really at home with things that shine and glitter. The Westerner uses silver and steel and nickel tableware and polishes it to a fine brilliance, but we object to the practice. While we do sometimes indeed use silver for tea kettles, decanters, or saké cups, we prefer not to polish it. On the contrary, we begin to enjoy it only when the luster has worn off, when it has begun to take on a dark, smoky patina. Almost every householder has had to scold an insensitive maid who has polished away the tarnish so patiently waited for.⁶⁶

It is the same for Didi-Huberman in this regard (see below, p. 156): a rejection of strong, harsh, powerful light in favor of quiet, subdued light – the nightlight, the firefly. Veiled light. A friend of mine has just the right word to describe this aesthetics and ethics of the shadow: *modesty*. Euphemism. It is the world of the folding screen as opposed to that of the lamp.

The folding screen tries hard to contain the moonlight;
the immodest lamp shines on a solitary sleep.⁶⁷

Lichen appears as a *baroque*, and thus *impolite*, organism. From the "irregular pearl" to the "worn stone" (lichen's acidity), tarnished, crudely soiled: "the effect of time."

> We are not at all biased a priori against everything that shines, but to a superficial and icy brilliance we have always preferred deep, slightly hazy reflections, be it in natural stones or artificial materials, that slightly altered shine that irresistibly evokes the effects of time. "Effects of time": it sounds good, but actually it is the sheen produced by the dirt of the hands. The Chinese have a word for it: "the hand's luster"; the Japanese call it "wear."[68]

Let us recall the designed stones, like the stone/landscapes (*suiseki*) of ancient China, so admired by Roger Caillois.

In the novels by Yasunari Kawabata (1899–1972), this poetry of the moss-covered garden and chiaroscuro is very subtly traced:

> The moment Chieko noticed the violets, several small white butterflies fluttered through the garden from the trunk near the flowers. The whiteness of their dance shone against the maple, which was just beginning to open its own small red leaf buds. The flowers and leaves of the two violets cast a faint shadow on the new green of the moss on the maple trunk. It was a cloudy, soft spring day.[69]

What captures the writer's attention in describing the beauty of this spring in the process of opening is precisely the effects of nuances: the veil of butterflies on the tree and the violets on the moss, the clouds, the tree, and the sky become blurred. It is all the effect of superimposition and blending, all the work of light cast and filtered. At night, the nuances of the garden are linked to the chiaroscuro provided by the moon and the colors of bodies reflecting this fragmentary gleam:

The Far East, Mosses, and Wabi-Sabi

It was a rather artless oblong garden, but about half of it was bathed in moonlight, so that even the stepping stones took on different colors in the light and shadow. A white azalea blooming in the shadows seemed to be floating. The scarlet maple near the veranda still had fresh young leaves though they were darkened by the night. [...] The garden also had a rich cover of hair moss.[70]

*

Wabi-sabi is rooted in the Taoism of ancient China, which highlights the passage of time, the wearing away of things, and a fascination with nature and its energies. What delicacy but also what power emanates from the old trees released from Wang Tingyun's ink brush in the twelfth century, as they appear flanked with lichens like so many deer herds!

Let us now aim the spotlight, or rather let us direct our small candle toward the lichens represented in Far Eastern poetry and painting.

Deer Park
No one seen. Among empty mountains,
hints of drifting voice, faint, no more.

Entering these deep woods, late sunlight
flares on green moss again, and rises.[71]

In Japan, lichen is part of the physical garden but also the poetic garden: the haiku. The French alexandrine and the French garden stand in stark contrast to the Japanese haiku and the Japanese garden (as well as bonsai). These are rhetorical and symbolic spaces, not seeking to domesticate but rather to reveal, to idealize the beauty of nature.

Nothing is moving
except the summer sky
lichen on the pines.[72]

The permanence of lichen. On the pines: symbols of the immutable. This poetic form requires the presence of a key word connoting, sometimes very subtly, one of the four seasons: the *kigo* (季語). Poets thus made use of a kind of almanac, called the *saijiki*, that collected the various words associated with the seasons. Lichen and moss were season words for spring and summer. The key word "lichen" symbolizes new green growth – even in tragic and apocalyptic circumstances.

> Earth burned by the A bomb
> let it be brightened up
> by lichen flowers.[73]

In the context of Japan and the trauma at the end of the Second World War, colorful lichens appear as flowers of consolation and hope, an expression of survival amid disaster, on the devastated earth of Hiroshima. Flower of evil, that grows in the ruins or that perseveres. Even burned, nature remains worthy of contemplation.

Traditional Chinese and Japanese painting chose nature as subject long before the West did. This sensibility was tied to Taoism, which made nature the place of spirituality and something to venerate. The conception of *physis* was different, involving less of a dominating, oppositional relationship. Since the first millennium, painters have been painstaking in their representations of nature, taking the form of imposing panoramas or, at the other extreme, small corners, details: a tree, a branch, a flower, and so on.

Although lichen is absent in the history of Western painting until very recently, it has been a traditional motif throughout the centuries in China and Japan. It appears at two special moments: the Song (tenth century), with its taste for the bearded fruticose lichens hanging from hoary branches; then the Ming in China, and the Edo in Japan (sixteenth to seventeenth centuries), as a model adopted by the Kano school, with foliose lichens delicately dotting the trunks of cherry trees.

Bearded lichen is very often represented in paintings that evoke the passing of time, the relationship of the past to memory; it is part of the code of motifs for antiquity in painting. Imbued with Confucianism, the doctrine of stability, Far Eastern arts pay homage to the past and the experience of old age. The usnea's beard is an extension of the sage's beard. The Chinese painter Li Cheng (919–987) – whose name is fittingly close to his subject – sketched usneas falling from a crooked old tree near a stela, which a walker and his attendant have stopped to admire. Lichens thus contribute to the verticality of the image – which echoes the verticality of the monument, existing through the ages. As one of the Song's great landscape artists, Li Cheng often represents winter in realist paintings, the trees dead or alive. This theme of the old tree became a veritable *topos*, often with its cortege of usneas. Wang Tingyun (1151–1202) took up this motif again in his painting *Secluded Bamboo and Withered Tree* with rougher, more stylized brushwork, just as Ni Zan (1301–1374) did. Jaccottet writes:

> The bright foliage of the oaks in the loose stones, on the brittle branches covered with silvery lichen, what could be more beautiful? But isn't this the image of the aged, declining poet with his very last budding of words?[74]

In China, traditional painting was largely done by the literati and derived more from calligraphy than from arts and crafts. The brush and canvas were the continuation of writing, of ink and paper. Landscapes are often stylized, semi-abstract, which explains why lichen is very often simplified in them, reduced to calligraphy, suggested by a small dark stroke, called the "lichen point," or the "moss point" (*tai dian*) in Chinese pictorial nomenclature. The lines of the branches can also evoke ideograms, and their verticality can call up that of writing. Far Eastern art intrinsically links writing and image. Within these conditions of abstraction, nothing allows lichen to be distinguished anymore from moss, a lenticel, or some other irregularity in the bark. The goal is not realism but the stylization of the landscape, in search of graphic effect. Examples are the sketches of lichen done in the Ming, by Wang Qian in particular.

That being said, traditional painting also presents a descriptive, decorative side, blossoming out on partitions and fans. Beginning in sixteenth-century Japan, the Kano school developed the motif of the cherry tree (*Prunus serrulata*: *ume* in Japanese). It is one of the "four venerable" subjects portrayed since the earliest days of Chinese art, along with bamboo, orchids, and chrysanthemums, and is an object of true veneration in Japan because of its splendid spring flowering.[75] It is the first tree of the year to blossom, sometimes even before the snow melts. Like the wind called *haru-ichiban*, it is one of the signs announcing spring. Its branches can be drawn against the sky with abstract lines that recall Japanese calligraphy (as in prints by Hiroshige). Very often painters sketched foliose lichens, especially of the *Parmotrema* genus, on its trunk and branches. Kanō Sanraku (1559–1635) and his apprentice Kanō Sansetsu (1589–1651) were fond of old cherry trees in blossom; their paintings were admired by Michel Butor.[76] Undoubtedly the most famous and most spectacular work is by Kanō Sansetsu, *Old Plum* (1646), originally located in a temple in Kyoto. The tree extends from right to left in an impressive twisting, convulsive movement along the entire width of a five-meter screen composed of four sliding panels (ink, color, and gold leaf on paper). The lichen painted on the trunk, no doubt *Parmotrema tinctorum*, serves a decorative function in the same way as the buds and white blossoms with which it interacts (green with a white border, it echoes their white corollas with green centers). It appears like a kind of bark flower, only distinguishable from the color of the trunk by its slightly lighter contours. This motif is found again in China with Chen Hongshou (1598–1652; see Ill. 7), genre painter (pastoral) and calligrapher, and then in Japan with Tamara Chokuo (active between 1688 and 1704) and Ogata Kōrin (1658–1716) of the Kōrin or Rinpa school. The US philosopher and art historian

Ernest Fenollosa wrote in *Epochs of Chinese and Japanese Art* (1912): "It is in this sense that we can call the chief masters of this Kōrin school the greatest painters of trees and flower forms that the world has ever seen." In Ogata Kōrin's striking *Red and White Plum Blossoms* (circa 1700–1716), the two trees and river shine radiantly in green, white, and red on gold leaf, without perspective. In contrast to the abstraction of the river at the center is the figurative precision of the blossoms and lichens on the two plum trees. On the trunk we again find these ornaments of foliose lichens (*Parmotrema*) whose bright color, extended by the areas of a few mosses or leprous lichens on the bark, contrast, to the right, with the complementary red of the buds and flowers. That is because in Japanese painting, lichens *are flowers*; aesthetically they are flowers in the same way as plum blossoms are. This painting brings to mind for me the refinement and contrasts in certain expressionist canvases by Gustav Klimt.[77]

The painters of the Rinpa schools – Fukae Roshū (1699–1757), Sakai Hōitsu (1761–1829), Kiitsu Suzuki (1796–1858), and Mori Sosen (1747–1821) – turned to the technique of *tarashikomi* to suggest on their folding screens the textures of tree trunks, leaves, and lichens. The artists applied a layer of color (or drops of ink, water, or pigments) on a surface already painted with water and still moist, in order to create a blurred effect (that "disturbance" in things) and a vibrant tone. Since it was impossible to know how the pigment or ink would spread after being applied on the moist area, the painter gave himself over to the whims of the material, but not excessively so. The result is integrated, subtly sensual textural effects that contrast sharply with the extreme clarity and finesse of the flowers and animals represented on the same canvas.

It is difficult to define a neat symbolism in the representation of lichen in the Far East. If the cherry tree can express rebirth after winter, the exaltation of life and survival, or conversely, the ephemeral nature of its flowering, then lichen must be seen primarily as a natural element among other natural elements, an object of veneration and decoration, with qualities of form, color, texture, and rhythm that are revealed in association, in symbiosis, with the cherry tree and other features of the landscape.

*

Another striking work by Ogata Kōrin shows the Chinese sage Taigong Wang – deified by Taoism – as a hermit, seated on the shores of the Isui river, shortly before he became a minister for the first Zhou kings, living an exceptionally long life. In the grass and on the bank can be seen green forms accentuated with white, which resemble lichens (possibly of the *Xanthoparmelia* genus, which grow on rocks along rivers). They echo the motifs in the cloth on which the sage is sitting and are thus associated with wisdom and longevity. That is why the traditional robes of Zen and Buddhist monks, known as the *koromo* or *goromo* [ころも] and worn over a white kimono, are sometimes called "moss robes" (*koke no koromo/ koke goromo*). Perhaps because of the color, certainly because moss evokes the condition of the hermit: solitude, simplicity, eternity (*koke no iori*, "moss hermitage," thus designates the cottage where he withdrew in the mountains). More generally, nature is the place of the sacred in Japan, and it is venerated as such; there is no rupture between the human and the rest of the living world. This continuity is visible through the art of the kimono which, as we have just seen, celebrates natures and creates a play of echoes: the landscape is clothing and clothing is a landscape. In this poem, the garden penetrates the space of the monk's chamber:

> A court full of palm blossoms,
> the moss-lichens penetrating the idle chamber,
> from one to another talking has ceased,
> in the air floats a strange perfume.[78]

Conversely, the expression *koke goromo* is used in classical poetry to describe moss, as well as the natural finery that it can evoke through metaphor. The word *goromo* also refers to the spotted skin of a variety of koi carp admired in Japan.

Green moss worn like a robe,
pack of lies on the shoulders of rocks;
white clouds designed like a belt,
to encircle the flanks of the mountains![79]

Figure 10a Ogata Kōrin, *Tai Gong Wang*, early 18th century, screen in two panels, originally in color, 166.6 x 180.2 cm, © National Museum of Kyoto, Japan.

Figure 10b Collections of lichens in the herbaria of Geneva's Conservatory and Botanical Gardens, 2018, © Vincent Zonca.

Part 3
ECOPOETICS: LIFE FORCE AND RESISTANCE

"Lacking sun, learn how to ripen in the ice."
Henri Michaux, *Tent Posts*, 1978

"Lichen on the post survives."
Hans Magnus Enzensberger, *The Fury of Disappearance*, 1980

Ruderal

In botany, the plants called "ruderals" are those that grow spontaneously in spaces *unintentionally* altered by human presence and often left neglected: parking lots, lawns, sidewalks, traffic circles, roadsides and riverbanks, garbage dumps, residential and industrial ruins, compost and trash heaps, tunnels, urban wastelands, no-man's-lands, and so on. Lichen does not grow specifically or uniquely in these spaces that humans have unwittingly transformed, but it is found in our cities where it was never cultivated, in abandoned areas, on the margins of urban and rural landscapes. In this sense, the trunks of trees, the length of streets, the walls or cornices of neglected buildings, the stones of quays are so many urban "ruins" (or anthropic barometers – that is, for measuring human activity) in that they let grow spontaneous, disobedient, uncivilized organisms, just beyond control.

It is precisely a matter of proposing lichen as an alternative, peripheral place (sometimes precisely situated in urban centers themselves, under our feet, on the walls we walk along or the buildings we pass most often) for thought, inscribing it at the heart of a new relationship with our environment.

Lichen has much to teach us about our relationship to the environment and our place within the living world, within nature. At a key moment in our history, increasingly viewed as the "Anthropocene age" (henceforth humans have an irreversible and global effect on the planet, including its geology, which once seemed immutable and infinitely superior), this little organism with its incredible life force invites a new vision (on a new scale), a new ecology. It also expresses the evolution of ecological and ecopoetic discourse, in the West in particular.

Rousseauist Walks

> "I know no study more congenial to my natural inclination than that of plants; the life I have led for these ten years past, in the country, being little more than a continual herbalizing, though I must confess, without object, and without improvement."[1]

Jean-Jacques Rousseau (1712–1778) drew his poetics and his politics from the contemplation and study of nature. In Book V of the *Confessions*, he admits to his wholly personal penchant for botany. By the end of his life, he was madly in love: "I am wild about botany; it gets worse every day; my head is filled with nothing but hay; I am going to become a plant myself one of these mornings," he wrote on August 1, 1763 to his mentor, the Neuchâtel doctor and botanist Jean-Antoine d'Ivernois.

For him, botany was fundamentally a contemplative art practiced for its own sake. By studying bryophytes and other plants, he was not looking for their therapeutic or industrial properties, nor looking to create finery and decorations, but trying to know and contemplate the beauty of creation (the pursuit of knowledge had not yet yielded to the pursuit of power, a change Paul Valéry's *Essais* addressed in the early twentieth century). This was the science of nature "that does not lie," and that invites us to merge with it – note Rousseau's vivid remark about "becoming a plant." It is, in this sense, an ethical study: assenting to what is (an Edenic nature thus idealized, that is, the formulation of an ecological consciousness), seeking an openness to the world and the self, out of vexation with humanity. In 1764, Rousseau was living at his property in Môtiers in the canton of Neuchâtel, seeking refuge from condemnations brought against him and his books in Paris and Geneva, when he discovered the plant world. "The study of plants was the proving ground for Horace Bénédict de Saussure's nascent powers, as it was the diversion and consolation of Jean-Jacques Rousseau's later days," wrote the famous Genevan botanist Augustin Pyrame de Candolle.[2] It was a passion they shared with another Enlightenment figure, Malesherbes, with whom Rousseau corresponded and who was known to travel about France, Holland, and Switzerland incognito, under the pseudonym of "Monsieur Guillaume," returning with materials and observations on agriculture and industry, as well as plants for his collections.[3] This was a century with a thirst for knowledge and a passion for reason, for exhaustive encyclopedists, for collectors.

> They say a German once wrote a book on a lemon-skin; I could have written one about every grass in the meadows, every moss in the woods, every lichen covering the rocks – and I did not want to leave even one blade of grass or atom of vegetation without a full and detailed description.[4]

At the very time when he was composing *Reveries of the Solitary Walker* (1776–1778), Rousseau was collecting, botanizing, and popularizing. He traveled throughout the countryside, especially around Neuchâtel and in the fields and woods of the Jura in order to create elegant little herbaria intended for his own studies or for his friends. By the late eighteenth century, collecting and studying mosses had become a favorite pastime of the wealthy (a diversion for nobility, and then in the nineteenth century, a leisure activity for sons of rich bourgeois Genevan families). In sitting rooms, mosses became all the rage. Mosses gathered, mosses clung! Samples, writings, collections were exchanged. Mosses became a language, words sent back and forth.[5] Voltaire's *Dictionnaire philosophique portatif* corresponds to the little herbarium – or *moussier* – portable and meant for taking on walks, for botany lessons. Moss and lichen were part of Rousseau's cabinet of curiosities, calling for certain social practices ("Botany is my passion, and it is my pleasure to tell you about it"), as did his solitary walks in privileged harmony with nature. Thus for Rousseau, botanical engagement was entirely noble, a healthy, educational diversion; contemplating nature was more valuable than many frivolous activities, better than the decadence of culture.

> I think your idea of amusing the vivacity of your daughter a little, and exercising her attention upon such agreeable and varied objects as plants, is excellent; [...] convinced that, at all times of life, the study of nature abates the taste for frivolous amusements, prevents the tumult of the passions, and provides the mind with a nourishment which is salutary, by filling it with an object most worthy of its contemplations.[6]

I remember how I discovered, as a very young student visiting the *Jean-Jacques Rousseau et les arts* exhibition at the Panthéon in Paris in 2012, the incongruity of his "*moussier*" as he called it: his book of mosses.[7] This was, for me, the beginning of an intriguing mystery. A herbarium that contained no flowers or leaves, but patches of moss and scraps of lichens? A herbarium that was made not just of pages on which plants were attached, but envelopes and boxes containing the object – moss, detached or on a stone or branch – clearly wrested from nature, denatured. Rousseau mentions this difficulty in one of his *Lettres à Monsieur de M*****:

> Always alone and with no master other than nature, I have made incredible efforts with very little progress. By working very hard, I have managed to learn basically to determine the genera; but for the species, whose differences are often poorly marked by nature, and even more poorly described by authors, I have come to be able to distinguish with certainty only a very small number, particularly in the family of mosses, among the difficult genera like the *Hypnum, Jungermannia,* and *Lichens.* [...] But there is another difficulty; it is that mosses with their small strands do not catch one's eye at all on paper like they do on the ground, gathered into clumps or tightly packed turf. Thus it is useless to study them in a herbarium and especially in a *moussier* if one does not begin by studying them on the ground.[8]

Further on however, he advises his friend not to get lost in the details of collecting and inventorying: good philosophers of the Enlightenment were concerned with generalizing, abstracting, and examining the small to better focus on the large.

The exploration and expansion of knowledge that characterized the Enlightenment was applied on the local level, within the cosmos of the village, which explains the shift in attitude toward lichen and other organisms. Everything was deemed worthy of being the object of study and art, especially the familiar. The principles of the Enlightenment could henceforth be applied on the street corner, in the vegetable garden or the drawing room. It was no longer necessary to travel to French Guiana (although one could certainly buy the herbaria of traveling naturalists, a pleasure Rousseau could not resist): "to my mind, the greatest charm of Botany is to be able to study and know the natural world around oneself rather than in the Indies." The great explorer was the one in the Swiss orchards.

Georges Perec has this word "endotic" to name, on a cultural level, the necessary reorientation of our regard toward everyday things around us:

> What's needed perhaps is finally to found our own anthropology, one that will speak about us, will look in ourselves for what for so long we've been pillaging from others. Not the exotic anymore, but the endotic. To question what seems so much a matter of course that we've forgotten its origins. [...]⁹

Rousseau invites us to linger over the small things that surround us, and that we are no longer used to looking at. Today, such humility within our globalized world, where distances are reduced and the exhilaration of exoticism and desire encouraged, offers food for thought, even as the experience of confinement during the pandemic in 2020 reshuffled the deck and invited us to relocalize our gaze. Ruderal: returning to the local, interest in the endotic and the everyday.

Botany takes on a personal dimension for Rousseau. And in *Reveries*, mosses are associated with images of laps, nests, hollows (Bachelard's "corners"), refuges for the paranoid or the misanthrope, and supports for poetic reveries and fantasies, poetic prose. The solitary walker is above all an ecologist, a botanist:

> Some small birds, scarce but familiar, tempered the horror of the solitude. There I found the notched *heptaphyllos*, the *cyclamen*, the *nidus avis*, the greater *laserpitium*, and a few other plants, which occupied and delighted me for a long time; but gradually succumbing to the powerful impression of my surroundings, I forgot about botany and plants, sat down on the pillows of *lycopodium* and mosses, and began dreaming to my heart's content, imagining that I was in a sanctuary unknown to the whole universe, a place where my persecutors could never find me out. Soon this reverie became tinged with a feeling of pride; I compared myself to those great explorers who discover

a desert island, and said complacently to myself, "Doubtless, I am the first mortal to set foot in this place." I considered myself well-nigh a second Columbus.[10]

The rational thinking of the Enlightenment was thus combined with an ecological awareness and sentimentality that, in Rousseau, took the form of an Edenic and solipsistic idealism. It was a matter of rediscovering the innocence of an original, pure, and lost nature. Culture was necessarily cut off from it; it was the other name for this loss. Mosses and lichens were, in all this, personal and philosophical objects of a fantasy, the dream of a return to the natural matrix, the bosom of the lost mother. Rousseau's ecology, with its obvious psychoanalytical and religious roots, did not yet allow for the conception of a world that would go beyond the nature–culture opposition.

*

While I was rediscovering these *Reveries*, which I had read for the first time in high school, I was also walking Lake Geneva's magnificent Shore Path that goes from the Pâquis district to the Place des Nations. Looking at the trees in the Botanical Garden, and the surrounding hillsides, I tried to imagine Rousseau a little more than two centuries earlier, busy botanizing while the French Revolution appeared on the horizon.

The encyclopedist's generalizing, it seems to me, intersects with the specificity of the Genevan mind. Geneva, the very city where I was living at that moment and where I began writing this book, played a revelatory and actuating role for me. It allowed me to better feel and think about this fascination for lichens; perhaps I would not have written it if I had not lived there, crossed through its old center of old walls, felt this worship of solid ground and austerity (with all the ambiguities of a city that is also ultraliberal, cosmopolitan, and ostentatious). Maybe it gave me less of a taste for order and regulation, for the pleasure of order (happily forgone in this book), than it did a sensibility more open to nature and small things, even in excessively flashy Geneva.

In fact, as the city of Saussure, Rousseau, the Candolles, Bonnet, and Gascar for a time, Geneva has been one of the great centers for botany since the nineteenth century.[11] This was especially because of the influence of Rousseau, the Enlightenment, and the pre-Romantics.[12] Various reasons can explain the close ties between this place and this science – and the sciences in general.[13] The very early development of the economy and the middle class allowed the best scholars to find ideal conditions there for pursuing their research and allowed the sons of great banking families to establish themselves and their servants there while pursuing their studies in Paris, thus linking the cities in matters of botany. "All the botanists, biologists, geologists, and entomologists, or nearly all, whose names decorate the university districts, came from the families of bankers."[14] Economic development found favorable conditions in the Protestant religion. Moreover, because of this same enlightened Protestantism, "scientific observation was not linked [...] as it was in the case of the French encyclopedists, to religious skepticism."[15] With the two spheres of science and religion being separate, the sciences could advance without restraint.

The Protestant ethic found in the familiar and perceptible, in simplicity and humility, in the study of the perfection of divine creation (nature), a form of security and wisdom that we feel permeating Rousseau. "The rational spirit of Calvinism [...] leads to method and a predisposition for rigorous thinking," writes Pierre Gascar.[16] It merges faith with knowledge and instruction. Gascar, who lived for many years in Geneva when he worked for the UNO, has reflected at length on Genevan "pious humility" and moderation:

> I have always admired the taste for moderation here, for limited possessions, this worship of the everyday that aims at reducing the world down to its proper size, and that led Geneva residents

to domesticate the edge of what is already called, significantly, the "little lake," by means of a water jet, which, by restricting their view, is meant to ward off a horizon too distant and too vast.[17]

Even more scathingly, Nicolas Bouvier writes:

Evidently, Geneva could not be like Venice, or later, Vienna, a city of unbridled passions and romantic love. In Geneva, Venus was put to death first by Calvinist misogyny, then by Victorian prudery. [...] There is nothing reprehensible ethically, morally, or sexually in being interested in fossils, pollen, orchids, or coleoptera.[18]

*

For Rousseau, to contemplate nature was to contemplate the work of God, a vision shared by the European botanists of the nineteenth century, and also with the English painter and writer John Ruskin in 1885.[19] To those for whom humility is nobility, lichen and moss are Christian:

We have found beauty in the tree yielding fruit and in the herb yielding seed. How of the herb yielding *no* seed, the fruitless, flowerless lichen of the rock? Lichen, and mosses (though these last in their luxuriance are deep and rich as herbage, yet both for the most part humblest of the green things that live), how of these? Meek creatures! The first mercy of the earth, veiling with hushed softness its dintless rocks; creatures full of pity, covering with strange and tender honor the scarred disgrace of ruin, – laying quiet finger on trembling stones, to teach them rest. No words, that I know of, will say what these mosses are. None are delicate enough, none perfect enough, none rich enough. How is one to tell of the rounded bosses of furred and beaming green, – the starred divisions of rubied blooms, fine-filmes, as if the Rock Spirits could spin porphyry as we do glass.[20]

The Protestant mind: favorable substrate for botanical studies. How to explain otherwise why, even today, botany is most advanced in countries like Switzerland, Germany, Sweden (home of Carl Linnaeus and Erik Acharius), and Finland (home of Wilhelm Nylander and Edvard Vainio)? Moreover, the word for moss in French (*mousse*) comes from old Scandinavian. [...]

*

In Helsinki I was struck with this "pious humility" that informs even the city's architecture. Faded colors like those of dry lichen before rain. The concern for moderation and detail. The close relationship to the environment that makes weekend and vacation retreats to the forests the favorite pastime. And what a surprise to find, in the botanical garden of the University of Helsinki, the world's first "lichen garden"![21] In Finland, as in Sweden, Norway, and Russia, it is part of daily life. It is a main source of nutrition for the reindeer (reindeer lichen creates a veritable winter meadow across the ground) if not for the humans surviving so close to the poles.

*

Romanticism expressed a new sensibility with regard to our relationship to nature. Following Rousseau, there were many Romantic artists who fell in love with botany and subjectivized that science. George Sand (1804–1876) adopted Rousseau as a model and gave detailed observation, as well as daydreaming, highest priority: "I dream, therefore I see"![22] She delighted in "bathing in botany," and in taking walks to collect plants along the Indre River.[23] Accompanied by her faithful friend the botanist Jules Néraud (1795–1855), who introduced her to botanizing, she set out in search of insects, stones, and plants:

> I remember one autumn that was devoted entirely to the study of mushrooms and another autumn that was not long enough for the study of mosses and lichens. For supplies we

had a magnifying glass, a book, a tin can to hold and preserve fresh plants, and in addition to all that, my son, a fine child of four years, who did not want to be separated from us, and who acquired there and retained a passion for natural history.[24]

Lichen was part of Sand's passion for nature and resonated with her interest in the small, the rejected, the proletarian. In her fiction, it becomes the subject of elaborate images that frequently create a frightening and gloomy but also bewitching atmosphere; lichen is one of the sublime Romantic images.

> A vegetation appropriate to the grotto – huge lichens, rough as dragon's scales; festoons of heavy-leaved scolopendra, tufts of young cypress recently planted in the middle of the enclosure on little heaps of artificial soil, not unlike graves – gave the place a terrific and sombre aspect which deeply impressed Consuelo.[25]

This comparison with scales is striking, and already present among scientists studying "squamulous" lichens. In lichen is crystalized a fascination with the terrifying beauty of nature and its vital energy, residing in the smallest of organisms. Ruderal: the place of minimal resistance.

> Oh yes! Strength! That is the duty, that is the revelation of Sinai, that is the secret of the prophets! A longing for strength is the need for development that necessity inflicts on all beings. Each thing wants to be because it must be. [...] Oh Father! [...] See in everything the harshness of invasion, the obstinacy of resistance! As lichen seeking to devour stone! As ivy strangling trees, and powerless to break through their bark, winding around like an asp in a fury![26]

*

Philosopher, naturalist, and poet alike, a versatility destined to disappear with modernity and the specialization of knowledge,

Henry David Thoreau (1817–1862) developed a "transcendentalist" approach to nature in the United States, linked to a Protestant background and a context peculiar to the United States (its Edenic nature?). Like Rousseau, he praised walks as a way to better commune with nature (in his essay "Walking," published in 1862). Today he is considered one of the founders of a certain kind of environmental thinking. Not surprising, then, to find him completely fascinated with botany as well. Although his knowledge regarding lichens may have been rudimentary and his interest selective, he made many botanical observations of them, particularly during the last ten years of his life.[27] These were recorded in the seven thousand pages of his *Journal* (1837–1861), a magnificent work written from field notes that provides access to the most personal and original part of his thinking. There he gives poetic descriptions of "lichen days" in the Massachusetts woods, those cold, misty, melancholy days that reveal the brilliant presence of lichens.

Those moist days are paradise for lichens, synonymous for them with a return to life; they soak up the moisture, drink it in like lips, like seaweed rediscovering the sea. The droplets and diffuse light bring out their colors and textures again. I remember my surprise one morning, during the rainy, gray weeks of Genevan spring, to find the *Xanthoria parietina* transfigured on the trees because of the ambient moisture. From a vague yellow, they had become green, an unreal Martian green, all phosphorescent and rubbery! Thoreau was fascinated by this metamorphosis:

> A mild, misty day. The red oaks about Billington Sea fringed with usneas, which in this damp air appear in perfection. The trunks and main stems of the trees have, as it were, suddenly leaved out in the winter, – a very lively light green, – and these ringlets and ends of usnea are so expanded and puffed out with light and life, with their reddish or rosaceous fruit, it is a true lichen day. They take the place of leaves in winter.[28]

The winter mist reveals to the eye these lichens, transformed by the moisture and glowing with their fire on the tree trunks and branches, paradoxically revealing their extreme vitality at the harshest time of the year, the heart of winter, when other organisms have disappeared or lost their colors.

> A thick fog. The trees and woods look well through it. You are inclined to walk in the woods for objects. They are draped with mist, and you hear the sound of it dripping from them. It is a lichen day. Not a bit of rotten wood lies on the dead leaves but it is covered with fresh, green cup lichens, etc., etc. All the world seems a great lichen and to grow like one to-day [*sic*], – a sudden humid growth.[29]

Winter, the season when the world amounts to lichen. In this context, it is the only visible, growing being. The large surfaces of epiphytic lichens allow them to trap the fog and the nutrients found in it (ammonium nitrate, especially).

> There is a low mist in the woods. It is a good day to study lichens. The view is so confined it compels your attention to near objects, and the white background reveals the disks of lichens distinctly. They appear more loose, flowing, expanding, flattened out, the colors brighter for the damp. The round greenish-yellow lichens on the white pines loom through the mist (or are seen dimly) like shields whose devices you would fain read. [...] This is their solstice, and your eyes run swiftly through the mist to these things only. On every fallen twig, even, that has lain under the snows, as well as on the trees, they appear erect and now first to have attained their full expansion. Nature has a day for each of her creatures, her creations. To-day is an exhibition of lichens at Forest Hall. [...] Ah, beautiful is decay! True, as Thales said, the world was made out of water. That is the principle of all things.[30]

Revitalized by the fog and by this touch of humor, lichens give new beauty to the wet, washed-out, decaying landscape where everything is in motion, fluid, melancholy, that instability of bodies and elements.

> I find myself inspecting little granules, as it were, on the bark of trees, little shields of apothecia springing from a thallus, such is the mood of my mind, and I call it studying lichens. That is merely the prospect which is afforded me. It is short commons and innutritious. Surely I might take wider views. The habit of looking at things microscopically, as the lichens on the trees and rocks, really prevents my seeing aught else in a walk. Would it not be noble to study the shield of the sun on the thallus of the sky, cerulean, which scatters its infinite sporules of light through the universe? To the lichenist is not the shield (or rather the apothecium) of a lichen disproportionately large compared with the universe?[31]

Once again, lichen is compared to a winter star. "Puffed out with light and life" in the first excerpt, and then a star experiencing its "solstice" and "eclipsing" the trees, it takes the place here of the sun itself, its spores being the rays. We can well imagine a *Xanthoria parietina*, its gold disk shining, radiating like the sun in the moist winter light (Oscar Furbacken spoke of a "psychedelic sun"). If Thoreau returns to the "I" in this passage, it is because he identifies with and identifies his occupation with what he is looking at: he is a man who desires little, who lives on nothing, whose ethic is grounded in the dazzling but modest life of lichen. Hence the forest becomes the starry web of a new sky for the one who knows how to linger there.

> It is a lichen day, with a little moist snow falling. The great green lungwort lichen shows now on the oaks, – strange that there should be none on the pines close by, – and the fresh bright chestnut fruit of other kinds, glistening with moisture,

brings life and immortality to light. That side of the trunk on which the lichens are thickest is the side on which the snow lodges in long ridges.³²

This power of lichens to reveal life and immortality at the very spot where it is most difficult – the snow, the thaw – is for Thoreau the walker an impressive sight, incredibly powerful, that he literally devours with his eyes. The cold and damp paradoxically become the best of tonics. It is an extremely strong poetic image, at the intersection of poetry and visual arts. Looking at lichens allows a privileged relationship with nature to form, by licking the life force from it. In this sense, lichenology is a veritable ethic, an organic diet for Thoreau. If he takes such an interest in lichens, it is because they allow him to form the closest relationship to the earth, to have an ascetic experience. To see: to learn to live on nothing. To see: to touch and absorb what one sees. From vision we move to ingestion here, to the point of gradually merging lichenologist and lichen, subject and object:

> Going along the Nut Meadow or Jimmy Miles road, when I see the sulphur lichens on the rails brightening with the moisture I feel like studying them again as a relisher or tonic, to make life go down and digest well. [...] That's the true use of the study of lichens. I expect that the lichenist will have the keenest relish for Nature in her every-day mood and dress. [...] To study lichens is to get a taste of earth and health, to go gnawing the rails and rocks. [...] The lichenist extracts nutriment from the very crust of the earth. A taste for this study is an evidence of titanic health, a sane earthiness. [...] It fits a man to deal with the barrenest and rockiest experience. A little moisture, a fog, or rain, or melted snow makes his wilderness to blossom like the rose. [...] A lichenist fats where others starve. His provender never fails.³³

In these lines, lichen ensures a direct link to the earth. It is an existential diet, a minimal experience of life, but fortifying and healthy. It allows Thoreau to rediscover a wild energy, indispensable for the writing that he links intrinsically to nature and not to culture (to which he opposes it). The following lines have become famous, drawing up the manifesto for a wild literature:

> In literature it is only the wild that attracts us. [...] It is the untamed, uncivilized, free, and wild thinking in *Hamlet*, in the *Iliad*, and in all scriptures and mythologies that delights us – not learned in schools, not refined and polished by art. A truly good book is something as wildly natural and primitive, mysterious and marvellous [sic], ambrosial and fertile, as a fungus or a lichen.[34]

An organism that humans neither know nor plant nor domesticate and that establishes itself naturally, lichen is described here in Romantic terms. Thoreau is extolling a wild nature that must above all be found in ourselves, a primitive, undisciplined, imagined nature, like a myth: rough and ruderal.

> The knowledge of an unlearned man is living and luxuriant like a forest, but covered with mosses and lichens and for the most part inaccessible and going to waste: the knowledge of the man of science is like timber collected in yards for public works, which still supports a green sprout here and there, but even this is liable to dry rot.[35]

*

In 1719, Erasmus Darwin (1731–1802), English poet and naturalist, grandfather of the famous Charles, described his love for plants in two works, *The Economy of Vegetation* and *The Loves of Plants*. The latter is a long epic poem in four cantos, structured around Linnaeus' revolutionary classification system for plants according to sexual characteristics, and interspersed with philosophical commentary. It attempts to conceptualize botanical science and make it more accessible by personifying plants and emphasizing their relationships with human beings. A short passage is devoted to lichen:

> Retiring LICHEN climbs the topmost stone,
> And 'mid the airy ocean dwells alone. [36]

According to an already very Romantic aesthetic, it is associated with solitude (Césaire will turn it into allegory: "Clothed in lichens and epiphytes/ Solitude which passes [...]"; the same is true for Leopardi's broom, as we will see below) and summits (Goethe: "From the height of this entirely naked summit from where I can barely distinguish a few lichens growing at its foot, I think how one is seized with such a strong feeling of solitude," *On Granite*, 1784). Erasmus Darwin insists upon its pioneer nature: "This plant is the first that vegetates on naked rocks, covering them with a kind of tapestry. [...] In this manner perhaps the whole earth has been gradually covered with vegetation. [...]" The same mythic image is repeated by the poet William Wordsworth (1770–1850) who had read *The Loves of Plants* and was open to such attentiveness. In his dramatic monologue, "The Thorn," he writes:

> Like rock or stone, it is o'ergrown,
> With lichens to the very top,
> And hung with heavy tufts of moss,
> A melancholy crop:[37]

In the midst of the first industrial revolution, Darwin was one of the very first to note the relationship between pollution and the growth of lichen; he mentions in *The Loves of Plants* that lichen communities are affected by pollution close to metal foundries on Anglesey Island in northern Wales as well as around the Parys Mountain copper mine. In the United Kingdom, lichen has been linked to "smog" (it is sensitive to sulfur dioxide); a great many studies on this organism would be conducted there in the 1950s.

Sentinel Species

It was in Paris in 1866, seventy-five years after Darwin's *The Loves of Plants*, that Finnish botanist and lichenologist Wilhelm Nylander (1822–1899) was able to demonstrate the link between atmospheric pollution and the growth of lichens.

In his work, he observed that trees in central Paris were less covered with lichens than those in the woods on the outskirts of the city. He also showed that lichens gradually deserted the capital between 1854 and 1896. That is, at the time of *Fleurs du Mal*, the modern city (and "modernized" gardens, wrote Nylander), the works of Baron Haussmann and tall buildings, the second industrial revolution.

> The chestnut trees along the Allée de l'Observatoire are especially remarkable there because of the many lichens that cover their bark, in such abundance that one must go outside the city to find anything like it. That circumstance certainly lets us affirm that the part of Luxembourg of which we are speaking is the healthiest place in all of Paris. [...] Of all vegetation, lichens are the ones most widespread in nature; they live on bark, wood, rocks, stone, earth, everywhere that these substrates are bathed in pure air that circulates freely, because it is essentially on atmospheric elements these aerophilic cryptogams are nourished. However most lichens seem to flee the cities, and those that one does encounter there are often only partially developed, in a sorediferous state or completely sterile. [...] Lichens constitute *a kind of very sensitive hygiometer*.[38]

Beginning with a very precise mapping of lichen populations on the trees and walls of the city, it was possible henceforth to measure pollution. Nylander's demonstrations revolutionized ecological practices: this small, insignificant organism allowed air quality to be evaluated ("hygiometer" comes from "hygiene"). For the first time, we had "bio-indicators" at our disposal, and even today, the mapping of lichens is used in conjunction with direct methods of physiochemical measurement for biosurveillance. Lichens are called "sentinel species" by scientists – "vigilant vegetation" writes the novelist Pierre Gascar.[39]

Unlike so-called "superior" plants (those that possess stems and proudly rise toward the sky) and fungi, lichens have no protective *cuticle*, that external, impermeable layer provided with openings called *stomata* that regulate and filter exchanges with the atmosphere. Since they have no root system and draw nothing from their substrates (higher plants feed on oligoelements present in the soil), lichens, as aerial organisms, turn entirely toward the air. In 1754, Genevan botanist Charles Bonnet wrote: "Plants are planted in the air, almost like they are planted in the earth." Vitality but also vulnerability. Air quality has a powerful influence on their development; like a sponge or a sensor, lichenic substances store all the environment's micro-particles, atmospheric pollutants (lead, fluorine, heavy metals) as well as the elements that are essential to them (water, light, CO_2, and so on). In this sense, rather than a *tongue*, lichen might be a *mouth*, wide open to what the atmosphere offers, unfiltered. Its immersion into the fluid matter of the world is absolute. Sometimes merging with the mineral world, lichens are the most aerian of organisms, they are breaths. They provide stones and tree bark with a second skin, but they themselves are often "skinned."

Every species is more or less sensitive to these micro-particles; for example, some cannot tolerate sulfur dioxide (SO_2) and acidity, typical pollutants from Nylander's era until the twentieth century. Others are "nitrophilic," that is, partial to nitrogen (NO_2), like *Xanthoria parietina*; its presence, very frequent today, signals the nitrogen pollution characteristic of our time. *Xanthoria* is thus, in the strict sense, a ruderal lichen, because it develops as an unintentional result of human activity, in "abandoned" places. Usneas, which are much greater in volume, are more sensitive and much less resistant to pollution in general; these frayed beards have disappeared in our cities. The magnificent red lichen of the Americas, *Herpothallon rubrocinctum*, is also very sensitive to pollution.

In the city of Lille, because of industrial development, twenty-one species of lichens were recorded in 1901; in 1950, only two, in 1973, none at all.

Lichens are commonly used today to measure pollution in specific places and to identify the nature of pollutants: the fungi of lichens are very good bioaccumulators. They easily store heavy metals and radioactive substances, and can thus be used for sampling and measuring accumulated pollutants (the technique of biomonitoring); as bioindicators, they also make it possible to map their zones of distribution and their evolution in a given place.

That was the case, for example, during the Fukushima disaster in Japan in March 2011. Lichens of the *Usnea*, *Bryoria*, and *Alectoria* genera were used to measure the changes in the rates of cesium 134 and 137 in the air in Japan and far eastern Russia (the islands of Sakhaline and Kunashiri). In March 2015, the RMIT art gallery in Melbourne exhibited the work of six contemporary Japanese artists brought together under this title: "Celebrating the Opening of Japanese Art after Fukushima: The Return of Godzilla." Yutaka Kobayashi, born in Tokyo and a supporter of "green art," presented an installation there: *Absorption Ripples – Melt down melt away*. It evokes the speed with which a population can forget major catastrophes. Kobayashi reproduced a Japanese garden in the middle of which one rock shaped like Japan was covered with lichens and zeolite, a crystal also known for its ability to absorb heavy metals and radioactive elements, in order to show that if, as the news tells us daily, it is rapidly fading from Japanese consciousness, radioactivity itself will persist on the island for millennia. Zen garden or ground zero.

For several years now, a new technique developed in Finland notably, allows for the use of lichens to measure the humidity of forests, and thus to study the consequences of deforestation and the destruction of ancient forests.

Ill. 1 Anton Elfinger, medical illustration (lithograph) of *Lichen ruber pilaris*, in Ferdinand Hebra, *Atlas der Hautkrankheiten* [*Atlas of Skin Diseases*], published by the Imperial Academy of Sciences (Vienna: Braumüller, 1856-1876). (© F. Marin, P. Simon / Bibliothèque Henri Feulard, Hôpital Saint-Louis AP-HP, Paris).

Ill. 2. *Herpothallon rubrocintum* lichen, Brazil, Botanical Garden of São Paulo, 2018 (© Vincent Zonca).

Ill. 3. Crustoce and foliose lichens growing indiscriminately on the trunks of bamb
and a plastic chair, Brazil, Ubatuba, 2020 (© Vincent Zonca).

Ill. 4. Graphidaceae lichen of the *Phaeographina* genus, Colombian Amazon, Ticu
region, Rio Amacayacu, 2018 (© Vincent Zonca).

Ill. 5. Antoni Pixtot, *Saint George*, 1976, oil on canvas, 194.5 x 97 cm. (© Dalí Theatre-Museum/ © Antoni Pixtot 2022).

Ill. 6. Foliose lichens grow by rising through, by "licking," the surface of their support: here, a *Xanthoria parietina* on a deciduous tree branch, France, Burgundy, 2019 (© Vincent Zonca).

Ill. 7. Chen Hongshou, *Plum tree in blossom*, hanging scroll, ink and color on silk, 124.3 x 49.6 cm. (© National Palace Museum, Taipei).

Ill. 8. Leo Battistelli, *Globe Lichen,* detail, ceramic, 2020, exhibition at the Casa Roberto Marinho in Rio de Janiero, 2020 (© Fernanda Lins).

Ill. 9. Leo Battistelli, *Red Lichen 1*, 2019, ceramic, steel, and acrylic, exhibition at the Casa Roberto Marinho in Rio de Janiero, 2020 (© Fernanda Lins).

Ill. 10. © Oscar Furbacken, *Micro-habitat of Stockholm 1*, 2020, photographic print, 70 x 105cm

Ill. 11. © Oscar Furbacken, *Urban Lichen*, 2009, Sweden, Stockholm, Swedish Royal Academy of Beaux-Arts, photographic print on painted paper, 500 x 560 cm.

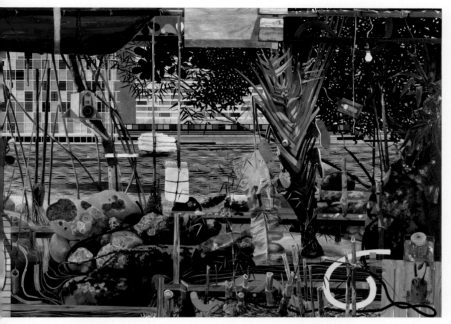

Ill. 12. Luiz Zerbini, *Mamanguã Reef*, acrylic on canvas, 293 x 417 cm. (© Eduardo Ortega).

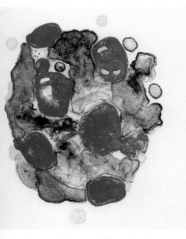

Ill. 13. © Yves Chaudouët, *?hens 1*, lithograph, Clot Studio, 1998. ADAGP, Paris and DACS, London 2022

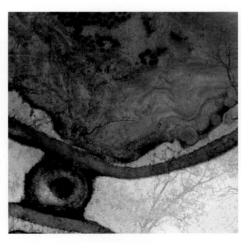

Ill. 14. © Pascale Gadon-González, *Cellulaires Contacts 5*, 2017, with Evernia prunastri (7296), lichen collected at La Vergne in Charente, 2017-2018, pigment print, 30 x 30 cm.

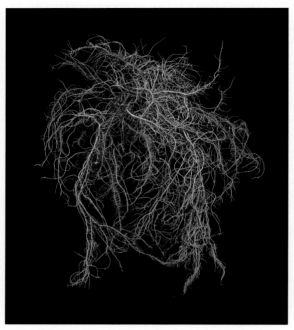

Ill. 15. © Pascale Gadon-González, *Bio-indicateur Usnea*, 1999, lichen collected in Corrèze à Maymac, lambda print mounted on Dibond, 120 x 80 cm.

Ill. 16. © Pascale Gadon-González, *Bio-indicateur Cladonia coccifera*, 1998, lichen collected in Ariège, lambda print mounted on Dibond, 120 x 80 cm.

Studies have been able to show that, although the industrial revolutions had the effect of eliminating most species from the cities beginning in early nineteenth century in Europe, reductions in the use of coal and developments in technology since the 1950s (with the use of particle filters, for example, in automobile exhaust systems, which let less sulfur dioxide escape) have allowed a very tentative return of lichens to urban environments. The air of our cities contains less sulfur than it did a few years ago, and certain lichens, especially crustose lichens, can tolerate it. Researchers at the University of Helsinki told me that in the 1970s and 1980s, the central botanical garden there was repopulated with lichens, as was Hyde Park in London. Walking around Paris and toward the Allée de l'Observatoire below the Luxembourg Garden, following Nylander's footsteps, I found the chestnut trees again flanked with foliose lichens, especially the *Xanthoria*, wedged in between the graffiti and posters.[40] Nevertheless, many genera like the usneas continue to become scarcer and disappear. It is estimated that twenty-five to forty percent of lichens will disappear from the earth's surface in the next sixty-five years, due to climate change and increasing development.

*

Despite the gradual return, in the West, of a few species of urban lichens, it is a whole ecosystem, of walls in particular, that is in the process of disappearing. The disappearance of old walls, the systematic use of herbicides and cleaning agents on tree trunks and façades, and the appearance of new pollutants all threaten the life of these organisms. Beyond all the romantic or poetic images, lichen is not only a precious indicator of life but also a fundamental element in the ecosystem that may disappear, watching over the quality of the air we breathe as well as over the urban ecosystem that surrounds us. The orderly and over-manicured nature of our modern cities is also a morbid nature of ornamental refinement, featuring trees and potted plants without considering organisms that could establish themselves spontaneously and participate in building an ecosystem, and then in the oh-so-precious humus. Recent city planning has tended to consider nature as *exterior* to the city, the city as an *abiotic* or *antibiotic* space, whereas the city itself is a natural ecosystem, natural, nonhuman, and unplanned as well as human and planned. It is also the whole concept of *patrimony* that must be revised and revitalized as well: architectural patrimony is coupled with natural, nonhuman patrimony, which also tends to disappear and must equally be displayed, protected, and left to overrun surfaces as much as possible.

I am thinking of the initiative undertaken by Geneva and its Conservatory and Botanical Gardens (Geneva – the city of the neat and civilized) aimed at protecting the old city walls through creating discovery trails. "This singular habitat, upon which many species depend, is often the target of inappropriate restoration and destruction. Useless cleanings, motivated by a false ethic of "clean and tidy" also threaten the organisms on the walls," reads the operations manual. These historic walls are called "sanctuary walls," but they are not only of historical, or even aesthetic and sentimental, interest. They harbor all sorts of organisms: mosses, lichens, fungi, flowering plants, spiders, acarids, insects, lizards, birds, and so on. Thus protection need not be synonymous with museification; it is a matter of drawing attention to these "living walls" and letting the local populations interact. And, while I was climbing the Rampe de la Treille to arrive at the historical center of Geneva, I was struck by the profusion and density of the species proliferating on the wall that bordered it, covered with asphalt: more than one hundred forty-nine species of plants and animals had been counted there in 2010.

"Lichens of sunlight and mucus of azure"

In Western literature, I really only uncovered lichen, identified as such and not as moss, beginning in the Romantic period. It is often associated with a heroic imagination – one of solitude – or a nosographic or fantastic one – that of decadence. Victor Hugo describes it as the "leprosy of time" in *Les Misérables*, in the same category as mold and bird droppings. It enjoyed success in gothic literature and dark Romanticism, which adored ruins and the macabre, among other ruderal organisms (mistletoe, ivy, weeds, and so on) populating haunted castles, abandoned houses, and tombs, "the places we would not want to live" (Goethe).

With the industrial revolution and the "modern" world, embodied in the poetry especially of Walt Whitman's *Leaves of Grass* (1855) and Charles Baudelaire's *Les Fleurs de Mal* (1857), an aesthetic rupture rocked the West. Poets articulated the crisis between the call of the ideal and the prosaic modern world. From then on, with the emergence of this new, torn sensibility, lichen thrived and, like acid, allowed for Romantic beauty and transcendence to be eaten away. "Orpheus with his lyre, all dressed in lichens!" cries the Italian poet Gabriele D'Annunzio (*The Flame*, 1900). Like the statue of the god of Love brought down in the gardens of Verlaine (in the poem "L'Amour par terre") and of Zola ("In the middle of this gloomy spot a mutilated marble Cupid still remained standing, smiling beneath lichens which overspread his youthful nakedness, while the arm with which he had once held his bow lay low amongst the nettles,")[41] the heroes of poetry are ruined, worn out, vestiges of another time.

> Free, smoking, topped with violet fog,
> I who pierced the reddening sky like a wall,
> Bearing, delicious jam for good poets
> Lichens of sunlight and mucus of azure. [...][42]

In "The Drunken Boat" (1871), Rimbaud makes the drunkenness of his new poetics speak. Through two particularly powerful oppositions, romantic and religious transcendence (Hugo's or Lamartine's "sun" and "azure") is brutally reduced to the triviality of "lichens" and "mucus" (the stress of the alexandrine makes the "*mor*" in *morves* [mucus] ring out ["*mor*" as in *mort* – death]. The romantic sunset is degraded into a wall down which the metaphysical drips, falling from great heights and becoming encrusted in the form of lichens. Three years later, for Flaubert, lichen is what hampers the transcendent and enigmatic words of the Sphinx:

> My feet, since they have been outstretched, can move no more. Like a scurf, lichen has formed upon my jaw. By dint of long dreaming I have no longer aught to say.[43]

The decadent poets and naturalist novelists of the 1880s continued down this same path, now conferring on lichen a positive value as well: it is no longer a question of opposition but of positing a new, modern aesthetic. Lichen became "beautiful," in its very triviality, in the same way as electrical lines, machines, or the Trans-Siberian Railway. Rimbaud's "dandy," was gaining prestige. US author Gillian Kidd Osborne notes a similar development in Herman Melville's work.[44] This is true especially in the collection, *Weeds and Wildings, with a Rose or Two* (1891).[45] Melville writes that "decay is often the gardener" – "Ah, beautiful is decay!" exclaimed Thoreau in 1851.

> If you only knew what kind of trash
> Poems shamelessly grow in:
> Like weeds under the fence,
> Like crabgrass, dandelions.
>
> An angry shout, the smell of fresh tar,
> Mysterious mildew on the wall –
> And a poem begins sounding fervent, tender,
> Making us all joyful.
>
> Anna Akhmatova[46]

*

Since the 1960s, living nature itself has become the material and the object of certain artistic practices, notably in the land art, *arte povera*, and new realism movements (let us think of Rauschenberg and his "grass paintings" in 1953–1954). The process sometimes has a political agenda, seeking to express a new relationship with the environment and art through provocative works, mundane and anti-lyrical materials, and a rejection of cultural institutions. The exhibition space is transformed into a laboratory for scientific experiments (terrariums and other tables instead of paintings) or moves outside into natural spaces. Nor is the temporality of the artwork fixed; it can take place over the long term (the exhibited work is alive, changes, grows, or is altered with the passing time: the work as an artistic and biological process) or very briefly (the work is its trace: ephemeral, it vanishes according to external conditions).

Lichen has found a prime place in all this, representing a "poor" (the word is used by Sbarbaro and Emaz) or primitive material. Thus, Peter Hutchinson (born in the US in 1930) experimented with organic materials in 1968, such as mosses, seaweed, lichens, or mushrooms (a botanist by training, he accompanied John Cage on a mycology walk in New Jersey.)[47] He then became interested in molds, displaying the pop and eclectic colors that emerge there: "large areas of orange and blue-green with many filaments of white and gray mold with a little red."[48] German-Swiss artist Dieter Ross (1930–1998) also explored molds in the late 1960s, in a more pessimistic light, in large decaying paintings-terrariums whose organic elements rotted and molded in the museum, with gases and other emissions gradually clouding the glass walls of the "painting." Ruderal: molds. There are also the installations, composite images (*To Enter the Burned Earth*, 2015), and plant sculptures (*Silhouettes*, 2017) – and the meals – of the French artist Tiphaine Calmettes (born in 1988), who sees in lichens, mosses, and other ruderal plants signs of renewal, despite everything – as did Hutchinson with his molds, "symbolizing life, representing the emergence of the beginnings of life and the reuse of base material."[49]

"Sbarbarian" Glowworm

It was over the course of those "decadent" years that Camillo Sbarbaro (1888–1967) was born in the province of Genoa in Italy. Poet-lichenologist, lichenologist-poet, still as well known today among scientists as among literary types, he began writing at the time of the First World War. It was in trenches as a volunteer with the Red Cross that he started building a herbarium composed in particular of mosses and lichens – lichen, once again, linked to "critical" experiences. Like Rousseau, in thirty years of research, he discovered one hundred and twenty-seven new species – about twenty of which bear his name – and today he is considered to be the greatest Italian lichenologist of the first half of the twentieth century (the other great Italian lichenologist of that period was Eva Mameli Calvino, the first woman to hold a tenured university position in botany in Italy, and the mother of another famous writer, Italo Calvino). The story goes that in the small bedroom of his house in Spotorno, Sbarbaro had no books except one thick volume of lichen taxonomy, published in Uppsala, Sweden, in 1952.[50] His lichen collection was donated to various museums, among them the Giacoma Doria Natural History Museum in Genoa and the Field Museum of Natural History in Chicago.

> My love for lichens only wanes in two circumstances: when I am in love and when I am writing. Thus he perceived correctly, the one who, without knowing me, diagnosed in this passion a form of despair.[51]

Sbarbaro explained his increasing passion for lichens as an outgrowth of his psychological despair, an abundance of tragedy (his life was marked with many depressive episodes). A Pascalian diversion in order to mask the void? A Freudian obsession of the manic-depressive variety? This passion substituted for the passions of love and creativity when they were lacking. But his love for lichen did not necessarily wane during times of poetic activity;

rather there were conjunctions, alignments: Sbarbaro's writing *is* lichen. His scientific research and his poetic quest are joined.

This fascination for lichens and for nature in his native territory attests to a form of Romanticism and Rousseauism, but also to the evolution of lyricism in the early twentieth century, in the context of a disenchantment with the world.

A contemporary of Giuseppe Ungaretti, and born just before Eugenio Montale, Sbarbaro read the work of Giacomo Leopardi. For him, to write and to describe went hand in hand: his lichen collection corresponds to the inventory of the world in his poems. But it is a matter of an inventory on the smallest scale This is not the epic list of Neruda's *General Song*; it is the patient, meticulous construction of the memory of places through the infinitesimal. A general song, but a minimalist one.

> The room is cluttered, permeated with the odor of undergrowth because of a lichen herbarium. [...] Because to collect plants is to collect places. Nothing better retains the memory of a site than the plant that is itself born there; belonging to it, as the one that draws its nature from it and whose least characteristics are determined by it, it represents it in the most concrete way. [...] Dried out, it still suggests the manner in which the sun touched it.[52]

These lines describe all the poetry of herbaria and the fragments of places that they contain. The herbarium is not simply a collection of samples or a nomenclature; it is also evidence of the place where the sample was taken. Collecting plants and describing landscapes, in this sense, is the same gesture: collecting places. Lichens like poems are traces, shavings, metonymic fragments of the world. Many of Sbarbaro's poems are basically descriptions of landscapes, impressions of a moment.

> Epiphany of encounters. Like strata, I welcomed within the aspect of all plants [...], the imprint of a leaf, an elytron. In

the passionate inventory of a tiny part of the world, the one to which I have closest affinities, I unknowingly satisfied my "servile love of things."[53]

Sbarbaro's poetry is thus a way of opening, welcoming the world, attention to the small. It turned toward the exterior, toward things, with a love less linguistic (as in the poetry of Francis Ponge) than romantic. In this sense, to collect a new species of lichen and to name things in a new poem are one and the same act. These are discovered fragments of the world, a single "work in progress," the same "continuous song" as Herberto Helder would say. "Once again today, a new lichen: the world is not completed."[54] And there is the anguish that one day, everything may be discovered and named. Hence in some texts, this frenzy to evoke the poetry of the Latin or vernacular names for lichens. Lost in the world, the poet finds in the herbarium a form of compensation and hope of belonging. Epic and infinitesimal temptation to seize the world, which does not explain why *lichen* exactly.

This organism won his heart for one particular reason: its humble presence. "Later, drawn by my predilection for everything that exists in a muted way, I devoted myself to the most retiring life forms."[55] "Discreet."

> Lichens interest me as a neglected – poor? – life form. [...] Sympathy: the same that makes me approach anything that is not showy (people, landscapes), that for others is unimportant, impoverished.[56]

Lyricism in a minor key, romanticism *pianissimo* (the title of a 1914 poetry collection) and glowing weakly (like the glowworm, also called lampyris or the will o' the wisp, the title of a 1956 collection, *Fuochi fatui*). These sonorous and luminous images speak of the same quest for humility, a minimal resistance. Here we find again Dante's fireflies and what Georges Didi-Huberman recently said of them in *Survival of the Fireflies*: "erratic glow, certainly, but living glow, glow of desire and poetry incarnate." Because of this fragility, to name is to make exist, to make last:

> It is because the tree leads a life of plenitude and harmony incomparable to our own, and to give it a name amounts to limiting it; whereas if we greet the lichens by name when we encounter them, we seem to help them live.[57]

No pathos here, but an ethic, which takes the small for a model.

Lichen also fascinated Sbarbaro because of the power that resides in it. Its long existence offered him calm, a force against nostalgia. Lichen grows extremely slowly, a few microns per year for some species! Its capacity for resistance in hostile environments is no longer in question. "How they struggle for the least space!" Sbarbaro speaks of a "silent vitality," of a *vicenda eterna*: "what drew me to lichens, perhaps, was to have learned that we don't know what they are; but what is most moving about them is their great life force [*prepotenze di vita*]."[58] Like Nietzsche's "Übermensch," lichen seems to be for Sbarbaro a kind of "überorganism," perhaps? This same feeling is evident in Giono. In *Le Chant du monde* (1934), a novel written in homage to Walt Whitman and his *Leaves of Grass*, the healer Toussaint, responsible for unity between human beings and the world, thus evokes the vitality of lichen, merging microcosm and macrocosm:

> They drew near to the table. It also carried a load of stones and herbs like the table in the room into which Toussaint ushered his patients. [...] "Look," said Toussaint, "[...] Do you know what all these colored specks are? It's a small lichen, old as the world, living ever since the world was the world, still living, and not yet in bloom. One of our trees makes its own flower and sheds it again within four seasons. Just reckon a bit. Over thousands of years. What confidence! It's no bigger than a fly's hair, and it says: 'I've plenty of time.' Maybe if we looked at the world from

> on high, it would be the same, and we'd also say: 'What confidence! That's my game.'"
> He moved his soft finger on the small universe of lichens.[59]

Lichen, with its solitary, muted, resistant life, mirrors the poet and his writing. We could easily imagine Sbarbaro choosing this organism, to which he endlessly compares himself, for his self-portrait. The poetry collection is the "repertoire of our precarious beatitudes," from a "poor/life." The herbarium is a private journal. Each collected fragment is not detached from the man who names it. Each poem is the memory of the world but also of the poet, the projection of his happiness and his despair, the expression of an unstable existential equilibrium embodied by the precariousness of lichen. Lichen and despair: inextricably bound.

> Some mornings the feeling of existence hangs by such a thin thread that a movement of the head seems enough to descend into the abyss without it even tearing.
> But on the contrary, stubborn existence persists! How many times, before dying, we follow logic toward death!
> Today mine is a life on strike. Oh, let one drop settle into this harsh drought! Thus the soul invokes a breath of poetry.[60]

Stubbornness. And drought: the dehydrated poet. A landscape of lichen to express the existential crisis, the modern condition, and to embrace the dynamic of the elegy: affirmation of loss, then a cry: "alas!" or "oh!" – in the hope of recovery, of a (romantic) breath. But also to express solitude, cultivated (for example, Sbarbaro liked to walk in fir forests to feel the breath of life there).[61]

> Because of its misanthropy, the city is the only barrier that stops it. If despite everything, it enters there, it is to go take the air high in the church towers, or else to lose, with its health, its particular physiognomy. Infertile and morose, urban lichen suffocates. Human respiration pollutes it.[62]

This Rousseauist personification highlights the split between nature and culture. As we have seen (see above, p. 109), there are, in fact, some species of lichens that like human company. This image inevitably leads us to recall the tree of another Italian poet, Giacomo Leopardi (1798–1837): broom, the desert flower (and here we encounter another major *topos* of poetic inspiration), which also grows on "hardened lava" and offers consolation:

> Here on the dry flank
> of the terrifying mountain,
> Vesuvius the destroyer,
> which no other tree or flower gladdens,
> you spread your solitary thickets,
> scented broom,
> at home in wild places. And I've seen your shoots
> embellishing the lonely plane
> around the city. [...]
> Now I see you here again,
> lover of sad places that the world has left
> and constant friend of fallen greatness.
> These fields
> strewn with sterile ashes, blanketed
> by hardened lava. [...][63]

Leopardi extols the resistance of broom, evoking as well the ruderal zones surrounding the city (it is very resistant to temperatures as well as sterile soils; it is currently used to rehabilitate industrial zones). The last line of the poem ("your fragile immortal roots") is an eloquent – and transcendent – summary of this very paradox: broom, lichen, fragile, immortal beings, just like the poet, between precariousness and the absolute, between spleen and ideal.

Thus emerges this portrait of lichen as accursed poet ("lichens are *deracinated*, lichens are homeless")[64] and as Baudelairian flâneur (such influences recur frequently in Sbarbaro), even as Rousseauist walker.

> Sometimes, when I walk alone
> in the streets of the tumultuous city,
> I forget my destiny of being
> a man among men, and, like an amnesiac,
> torn from myself, I look at
> people with strange, open eyes.[65]

This form of life, solitary, modest, discreet, but also stubborn and powerful, corresponds well to Sbarbaro's words. Lichen is a familiar and beneficial presence; it becomes a friend, a double (*"vita fraterna"*), a "blessed love." "Thanks to lichens, there is not a single place where I feel alone, since there is no place, arid and desolate as it might be, that is not, for me, full of living presences."[66] We find the same image in Barry Lopez's work, regarding his Arctic expeditions. Expression of and antidote for solitude. The quest for lichen like the quest for a presence. Like a fantasy of belonging in the world, a maternal fusion, which is conveyed especially in the gesture of caressing the plant.

> When I feel my soul gripping
> each stone of the deaf city
> like a tree with all its roots,
> I smile an ineffable smile to myself, and
> raise my elbows as though trying to take wing.[67]

*

Seeking belonging, a link to the world. In the context of postmodern disillusion, facing a world void of spirituality and reduced to immanence, French painter Bernard Legay (born in 1956) also evokes the search for primal contact with the world through the image of lichen.

> After disillusionment, one looks at lichens, at stones, one tries to restore contact with the most elemental things. Thus if one chooses to live, he looks at what hangs on.[68]

His small format series *Adhérences* works precisely with the textures of crustose lichens, those that are the most securely fastened to their support, through the play of acrylic and polystyrene on the canvas. His paintings are inspired by the nature of the bocage of Lower Normandy and are self-reflective: the painting as the contact of the material on the canvas, as a medium of contact with the world; it becomes skin, lichen, "desquamation."

> To work on the skin of the painting, like lichen, it is a skin on things, to work on this skin, as what links us to the world, and at the same time, what separates us from it, protects us from it, to work on that surface.[69]

*

The voice of the poet can only be a lichen voice: from self-portrait to poetic symbol. Sbarbaro's writing is indeed constructed through this analogy, more indirectly than with Antoine Emaz. Sbarbaro's poetics want to be sober, stripped down and spare, with minimal, even "mineral" lyricism, as in petrification ("From where do I draw these words? From where does the petrified trunk draw the green of its last expulsion?"), as in lichenification: "am I in the process of mineralization?" he wonders.[70] He returns to the elementary stage of poetry, prior to the opposition between the subject that looks and the object that is looked at, in order to render the pure existence of the world, washed clean of symbolism – much like Guillevic in that way – and all song, the dry, pathetic condition of limp speech ("the mold is broken, what joy!"). With the use of *fragmenti* – "wood shavings" (*trucioli*) – he developed early on a modern poetics of discontinuity and brevity in a form sometimes close to aphorism.[71] Lichen as a form of thought in prose, in tatters: between "frayed thoughts" (Emaz) or "filiforms" (Sbarbaro), and other "tent posts" (Michaux). A tenuous thread. The thread: that of Giacometti and Emaz, that of our fate hanging in the hands of the Fates.

Aridity is the initial place of the Sbarbarian poetic experience. Existential aridity, the aridity of the blank page, the aridity of native ground, *locus amoenus*, sung in the poems: arid Liguria, *terre avare*, and the town of Spotorno. This was the mineral landscape onto which the existence of the poet was grafted: a "soul that took root in the stone of the town and would no longer know how to live elsewhere."[72] As early as 1914, this poet was announcing the end of Orpheus, of symbols and myths (D'Annunzio), the desolation of the modern condition: "She lost her voice/the mermaid of the world, and the world is a great/desert";[73] the loss or death to come ("the intimacy of black"). "Sbarbarian" "lichen-speech" is, as for Emaz, lyrical and critical experience, struggle, aridity, negativity: a "rescued line."[74]

Ecological Forewarnings

We are no longer in the great Walden forests, but in the broad-leaved woods of Pierre Gascar's (1916–1997) Périgord childhood, and then in Jura's slopes and chasms that marked him to the point that his literary work is permeated with a fascination for plants and nature, and especially for the relationship that human beings have with them.[75] This is true of his novels and short stories (*Les Chimères*, 1969; *Le Règne végétal*, 1981; *La Friche*, 1993), in those works that are closer to personal journals (*Pour le dire avec les fleurs*, 1988) or to eye witness accounts (*Le Temps des morts*, 1953, one of his first novels, recounting his experience in the Ukrainian Rawa Ruska prisoner of war camp), and in his biographies of important scientists (Buffon in 1983, Alexander von Humboldt in 1985, Pasteur in 1986, his preface to the 1989 edition of *L'Herbier des quatre saisons* by Basilius Besler, from the seventeenth century). Moreover, each of his works oscillates between novel, personal journal, and documentary journalism. A writer forgotten today, left to lie fallow, he is nevertheless the author of a singular, erudite, and imposing body of work, published in the "Blanche" and "L'Imaginaire" series by Gallimard from the 1950s to the 1980s.

Pierre Gascar, born Pierre Fournier, was one of those rare artists who are true lichenologists (or vice versa), like Camillo Sbarbaro who made it his occupation and whose works Gascar read, or Bernard Saby who spent hundreds of hours in the National Museum of Natural History in Paris, consulting books, plates, and herbaria, and with whom Gascar might almost have crossed paths, since he frequented the same museum in the years following the Second World War. He spent his time collecting plants and lichens in France as well as during his many travels.[76] At once sedentary and nomadic, he was always close to the populations he encountered.[77] He also enjoyed constructing "lichen scenes," visual compositions under glass, that in the end frustrated him and that he relegated to using as hygrometers or "clocks." His relationship with these organisms was intimate, sensual, even maternal.

> Contact with lichens has no equivalent in the plant world. It is so clean that it borders on coldness. Nothing of them remains attached to your skin. No down, no secretion, no powdery film covers them. Their surface is as smooth as rubber or the balloon used for certain hygiene devices (like condoms), which squeaks when you wrinkle it or run your finger over it.[78]

But contact remains disappointing, clinical. Moreover, he himself gives his lichenology practice a psychoanalytical reading. He links it to the search for an original belonging, a romantic fusion with nature – this man who lost his mother, a school teacher, at a very young age, after she was committed to an asylum, "as if my life, since my birth, had only been a rending":

> I began to collect lichens in order, I imagine, to make pass for scientific curiosity what was only a withdrawal into myself, I mean into the most archaic part of my being.[79]

Gascar's thinking was very influenced by the surrealism of the period, and his relationship to the sciences was also a matter of fantasy, symbolic reading, and imaginary projection. Witness his interest in certain painters and writers like Rimbaud or Nerval.[80]

Moreover, his writing owes much to poetry, in particular to Rimbaud's *Illuminations* and Nerval's *The Chimeras*, as well as to the realism of Flaubert, whose hallucinatory perception of reality he appreciated, "[...] a form of clairvoyance that appeared all the more supernatural for being brought to bear here on the banal, the quotidian, and thus giving an air of the miraculous to the most ordinary life."[81] Gascar was a great friend of Caillois, with whom he shared especially a fascination for the natural world, both mineralogical and botanical (he received the Roger-Caillois Prize in 1994 for the entirety of his work), and who helped him build his collection.[82]

Although he was also interested in algae (nostoc), corals, ferns, and other organisms connected to origins (and to supports), as well as herbs and flowers, lichens increasingly became his consuming passion. He relates this madness in *Le Présage*, a book he devoted to that organism. It takes on a slightly burlesque nature and is reminiscent of another surrealist writer from the same region, Matthieu Massagier (born in 1949). In this novel-essay, thought and dream converse in the same desire for ecological engagement. Lichen becomes the common thread for various travels throughout the world (across Siberia, Venice, the Indian state of Gurajet, Thailand, the Jura homeland). In these different anthropological spaces, lichen becomes, in different ways, the sign or symptom of degradation, the image for a discourse on decay. Published in 1972, before Chernobyl and Fukushima, this text comes in the very midst of the cold war, with the fear of nuclear escalation and multiplying nuclear tests, more or less secret and supervised (let us recall John Cage, of the same generation – his *Mushroom Book* appeared the same year, which was also the year that Greenpeace was founded). Certain apocalyptic tones and symbolic interpretations attest to a vision that is sometimes romantic, fantasized (often at the same time), and reductive: pollution, as we have seen, leads to the disappearance of some lichens but also to the development of other species; lichen's great antiquity must be qualified.[83] This sometimes tends to undermine the rigor of Gascar's thought and his conclusions.[84] Nevertheless, the ecological thinking – the forewarning – that he developed here is original (with much prophetic accuracy at times) and greatly benefits from its relationship to the imagination. If some of his poetic/symbolic descriptions may seem outdated, even a little sentimental, there are nonetheless spectacular, hallucinatory passages, especially on lichens, that account for this work's great power.

The five journeys Gascar writes about become so many ecological omens ("over the course of my travels, lichens cast me, sometimes with the violence of an alarm, into some of the dramas of today's world"). The point of departure for Gascar's thoughts is the recurring observation of lichens' gradual disappearance, whether in subarctic regions ("for several years the gradual disappearance of lichens has been observed in the boreal regions") or on the stones of Venice. In this period, in Russia and Sweden, the cladonias of the Far North were storing unprecedented quantities of radioactivity because of nuclear tests ("nearly a scientific method of espionage"); in Venice, the parmelias were abandoning the walls, sculptures, and wet woodlands of the Venetian Republic. But very quickly the scientific facts took on poetic and deliberately prophetic emphasis. From bioindicators of radioactive and industrial pollution itself (factories, transportation, tourism) to the development of post-war modernity, lichens became symbols, through the disappearance of species sensitive to certain elements, and announced a decline of civilization: "a slightly magic demonstration, a warning that was given to us not by some divine power, but by nature, the principle of all reason."[85] Moreover, Gascar takes up certain images pertinent to decadence: the city of Venice, lichen itself, leprosy, landscapes of ruins, and apocalyptic metaphors.[86]

In Sweden, reindeer flesh is now three hundred times more radioactive than that of bovids in the south of the country, and its consumption among the inhabitants of subarctic zones causes problems, the severity of which cannot yet be measured.

> Leukemias resulting from radioactivity sometimes develop later and genetic mutations only appear in the descendants of irradiated individuals.[87]

These "literal," corporal consequences (disease, like the fungi with Chernobyl, and the disruption of the lives of the peoples of the Far North, described by the author with an almost mythical circularity, through shrinking pasture lands for the nomads, an increasingly sedentary existence, and the change in diet for the reindeer) are thus coupled with a symbolic reading. With the disappearance of lichens comes the end of a civilization as well, that of the reindeer, which the author describes as the infancy of humanity and art (henceforth, the warming climate would involve abrupt changes in temperature in the Far North, resulting in the formation of an ice layer in the snowpack that reindeer aren't able to scratch through to reach lichens and other vegetation). Along with this, there is the ecological deterioration of the Venetian lagoon as well as the "gradual obliteration of Venice" under the guano from pigeons and the flocks of tourists.

> Lichens prove to be vulnerable only to abnormal changes in the environment. How not to admire the rightness, even the wisdom, of this reaction? How not to applaud this refusal, how not to see a lesson in it?[88]

Lichen as omen, but also as ethic. That is because, symbolically, lichen is also a trace of the origin of life. "By collecting lichen, I was peeling the back of the dragon-world." That is why lichen, individually pathetic, useless, and small, appears to be the bioindicator for a true collective apocalypse. Henceforth observation becomes impassioned, the occasion for visions:

> Does anything exist in the world more ill-formed, more suited to elicit despair than lichens beginning to wither and already decomposing? [...] An evil of our time, a consequence of our

knowledge and our pseudo-rationalism, a phenomenon entirely definable in terms of physics, the disappearance of lichens nevertheless took on the fantasmagorical nature of signs, of forewarnings, and unleashed something very close to an apparition in the polar night.[89]

The softness of their lines, as much as their colors, [...] for the cladonias [with glowing red points], makes lichens crepuscular plants. Arriving in the zone where they became abundant, to the point of forming a sort of elastic prairie, I had the impression of having reached Limbo.[90]

The disappearance of lichens inspires empathetic and dreamlike expression. But it may be that the lichens Gascar has detached and collected, now decomposing, inspire his most beautiful surrealist passages:

As soon as I had detached them from their supports, all the lichens tended to make up a dream flora. [...] Placed on a shelf, they rebelled, no doubt because of the dryness of the atmosphere, and raised on the sides of their thalli the wing-fins of ancient flying reptiles. The fructicose lichens bristled, lifting infinite bifid tongues like the prehensile stamens of certain sea anemones. By bringing all these lichens together, with no regard at all for composition, one creates, on the scale of an insect, a kind of antediluvian landscape or an ocean landscape at abyssal depths.[91] (Lichens sometimes resemble *Oneirophantes*, slugs living at great depths, covered with papillae and eyeless.)[92]

Here, the act of detaching lichen from its support is also one of distancing oneself from reality and plunging into the fabulous. In this description, the shifting perspective is very significant. Gascar goes from the "I" of the writer-lichenologist to the general and impersonal "one," an empathetic or demiurgic expansion at the moment when the perspective shifts suddenly to the scale of the "insect." Thus a new world appears: as in Bachelard's texts (lichen "*se mondifie*" – makes itself a world) and in Oscar Furbacken's videos (the change in scale reveals "the topographic map of an interior landscape"). An ecological metamorphosis somehow occurs by which the gaze is displaced, as well as effaced, abolished, to the scale of the lichen. The viewer is thus no longer the one who writes and looks, but very much the one who meets, eye-to-eye, the lichens or slugs. "What we see, which looks at us," Didi-Huberman would say. But with Gascar, this metamorphosis is a confrontation with a form of infinity. The gaze, now on the level of lichen, perceives in what it sees a mysterious power that exceeds it and that recalls to me the experience of the "aura" as Walter Benjamin theorized about it. The aura is this power in regarding the image, the feeling of being looked at in turn when one is looking, and of being dominated and possessed by this gaze. It is the palpable experience of alterity that borders on the inaccessible, a form of the sublime. For Benjamin, this could be the experience of a sacred image or a work of art or a scene from nature:

> The experience of the aura relies on conveying the manner, once habitual in human society, of reacting to the relationship between nature and man. The one who is regarded – or believes himself to be – raises his gaze, responds with a look. To have the experience of the aura – of a vision or a being – is to realize the capacity of the one who raises his eyes, or to respond with a look. [...] This look is dreaming, draws us into its dream. The aura is the appearance of a distance, no matter how close it is.[93]

Here, Gascar sketches out the fantasy of a reciprocal gaze: lichen releases an aura. Those eyes regarding us are the lichen's surveillance, how it addresses us, as bioindicator of the state of our world, as prophet, as "image of life," just as Gascar says below. The guilt-producing gaze, but also the protective gaze. Gascar makes lichen into a sacred, prophetic figure, aura, and oracle. Like an icon, it is a "vigilant vegetable" (he says), that watches over us and delivers a message that goes beyond us (the "forewarning").

*

Gascar read Sbarbaro extensively. He saw in Sbarbaro's lichenology an ethical model, a support for projections: retreats into the natural world and the modesty of lichen (to the point of merging, of becoming "man-lichen"), into his native Ligurian village, into fragmentary writing, were read by Gascar as Sbarbaro's quests for his lost tie to the world (an existential "feeling of not belonging"). The solitary inventory of nature also became a refuge in the face of a disappointing reality, in the face of mounting fascism in Italy at the beginning of the twentieth century, and the industrialization and globalization of the world at the end of that century.

But retreat is also hope: just as Dante follows Virgil to the depths of Hell, Gascar botanizes in Sbarbaro's footsteps. Retreat allows for the decentering necessary to ensure new foundations.

> And so I crossed fields, following Sbarbaro (an old Italian with the look of a child, wearing a round hat: I had seen his portrait in a book), in search of the very same lichens [...], with the feeling of approaching the secrets of the world and discovering reasons for hope there. Because to flee from the world is perhaps the wisest way of placing one's hope in it. It is to search for *the image of life* there where it flows back, concentrates anew, regathers strength.[94]

*

In one of the exhibition rooms of the Casa Roberto Marinho in Rio de Janeiro, I happened upon a series of pieces by Argentinian ceramist – and *carioca* for nearly twenty years – Leo Battistelli (born in 1972). Battistelli received me at his magnificent home in the heights of Rio, nestled right in the heart of those few incredible portions of the still preserved Brazilian Atlantic Forest. He had replanted some species of native trees there and built a clean and sustainable house, in symbiosis with the place. Everything is green, luxuriant, exuberant. In such a natural setting, no way to escape lichens, he tells me! They are everywhere, with all the abundance and excess of the tropics. If he has chosen to devote himself almost exclusively to lichens over the past twenty years, through photography and especially through ceramic sculptures and installations (see Ills. 8 and 9), it stems from his activism. It is an attempt to raise awareness and render lichens *visible*, and to show that their disappearance, due to pollution, is a bad sign. By enlarging them and running them along the walls of museum galleries ("Lichen for Art Institutions," The Latin American Art Museum of Buenos Aires, 2002), he hopes to reconnect the public with the surrounding nature, and to show that lichens are the "alarms for our own failings," signals of the degradation in our environment. The ceramic and porcelain work allows him to give their thalli an extremely hard consistency; it is a matter of maintaining their life force, rendering them immortal, and "crystallizing" them, "fossilizing" them, he tells me. Whether it is Argentinian lava, where certain species grow, or Brazilian clays, the material is worked at high temperatures (up to 1400° C) which creates a play of colors and transparencies (see Figures 6a and 6b above).

Fragility, Resistance

What captured Sbarbaro's and Gascar's attention in particular, to the point of finding resonance there on existential and political levels, is the paradox of an organism that is both extremely fragile and incredibly resistant: the precariousness and tenacity of lichen. It is a life force constructed precisely to counter an often hostile and changing external environment. It is the idealism of survival, the epiphany of having little, the hope and belief in modesty.

According to the LIFE experiment (*Lichens and Fungi Experiment*) done in 2014, two species of lichen taken from the Alps and Spain (*Rhizocarpon geographicum* and *Xanthoria elegans*) survived for a year and a half on the walls inside an international space station, and then started growing again upon their return to earth![95] They can withstand about one year of exposure to ultraviolet solar rays, cosmic rays, and extreme variations in temperature, that is, conditions similar to those found on the planet Mars.

These organisms can put themselves into a kind of stasis, an organic pause, while awaiting better living conditions. In 1880, the botanist Théodore-Polycarpe Brisson mentioned the idea of "an intermittent existence." I would say rather a double mode of existence, an alternative, in the sense that their existence continues during this watchful stage. Lichens are in fact capable of living in slow motion, in "dormancy" during droughts (hence their exceptional survival in deserts: in the Negev desert, one lichen of the *Ramalina* genus can withstand many consecutive years of dryness). They rehydrate and resume their biological activity (photosynthesis) as soon as there is sufficient moisture. Lichens can then become saturated and swell like a sponge, until they contain up to thirty times their weight in water. All their colors begin to shine, especially the greens of the algae. This process is called *reviviscence* or *anabiosis*.[96] This "pulsing" between dry and wet periods is the rhythm distinctive to the life of lichens. Hence their great capacity for adaptation – and reason as well to fantasize about their long lives, as the poet Jean Follain does here:

> Lichen lives on the fluids of the atmosphere alone, imposing a gentle and powerful melancholy. It can be considered immortal. If it has slept for many years in a herbarium, removed and rediscovering the air, it can refructify. It is hypothesized that the planet Mars supports no form of vegetation other than lichens.[97]

Some alpine lichens may be one thousand years old, and others, in Greenland, could well be over four thousand years old. Moreover, lichen can harbor in its thallus another champion survivor (species characterized as "extremophilic"): the tardigrade ("slow walker"), an animal that looks like a microscopic bear cub, capable of surviving at temperatures ranging from −272° C to +150° C by slowing its metabolic processes to 0.01 percent of their normal rate. In 1960, English writer John Wyndham published a novel entitled *Trouble with Lichen*, in which he imagines that a biochemist manages to use a rare lichen extract to slow down the aging process, allowing humans to live for more than two hundred years.

Contemporary "Poethics"

> "When you wake up into lucidity from the dark
> without courage
> to bear your life and the noise of apples that fall
> cling stubbornly to the lichens, to the mosses."
> Marie-Claire Bancquart, *Énigmatiques*, 1995

This fungus–alga is enjoying astonishing attention in the visual arts, as we have seen, as well as in the poetry of the last few decades.

Since the 1960s and the emergence in the West of a new pastoral poetry responsive to the world around it, it recurs throughout the existential and "poethic" – a word coined by Jean-Claude Pinson – quest of writers from different backgrounds but all demonstrating a new interest in even the most trivial living reality.

In Europe, lichen thus appears in the pages of poets as varied as Philippe Jaccottet, Jean Follain, Jaime Siles, Jacques Dupin, Guillevic, Tomas Tranströmer, Jacques Lacarrière, Olvido García Valdés, Hans Magnus Enzensberger, Jude Stéfan, and Joëlle Gardes.[98] In North America, it shows up in the work of Lorna Crozier, Joséphine Bacon, Ken Babstock, and Brenda Hillman. It can even serve as explicit metaphor for poetic speech:

"the lyre of lichen"	Nuno Júdice
"the word of lichen"	Olvido García Valdés
"poetry as lichen"	Antoine Emaz

Sometimes featured in the titles of the collections themselves (Nuno Júdice, *Lyre de lichen*, 1986; Antoine Emaz, *Lichen, lichen,*

2003, then *Lichen, encore*, 2009; Lucien Wasselin, *Stèles lichens*, 2012; Gérard Freitag, *Aurores des lichens*, 2016), it can also be evidence of a fundamental fascination for writing:

> I envy its persistence. In the larger world it asks nothing of anyone except a bit of stone or wood. You could say that it flowers badly, that it vegetates, but I love precisely its choice to be anchored, like seaweed in the open air. I like to think of poetry as lichen.
>
> Antoine Emaz

In 1999, a literary review was even created in Canada with the name of *Lichen*.[99] In 2016 in France, a contemporary poetry review took that same name. A simple thematic coincidence? A matter of fashion and cross-pollination, or the long awaited metaphor for postmodern writing? To what extent does it allow us to better question the historical context and "vegetable" imagination at the turn of this century?[100] To understand the emergence of a vital new relationship responsive to the world, a new ecology? What are the different – and new – elements that drive this metaphor? And is it possible to hear in it a shared voice among these different poets? The symbol of a lyricism of resistance or survival ("green stubbornness with a flower of bark," writes Lacarrière) and a ruderal imagination?

This image of slow time, which expresses the paradox of fragility and resistance, can be seen negatively, as a melancholy force, that of time passing and recurring memory, that of postmodern disenchantment and ruined landscapes gradually covered with lichens; but also positively, as a vital force that resists time and arid conditions. It seems that, since the 1970s and 1980s, "lichen voices" have emerged that are trying to establish themselves, no matter what, and that resist, "face north," confront inhospitable environments.[101] As a possible allegory for the essential precariousness of lyric speech, lichen attests especially to a particular context, a crisis of subject and poetics, of the present moment when the threatened voice persists, a "counter-poetry," both negative and positive in this sense, that is made through its unmaking, and is unrelenting.

Jaime Siles: Existential Landscapes and Gongorism

The Spanish poet Jaime Siles (born in 1951) began to write during the final years of Franco's dictatorship. He takes a stand through the very means of a neutral, abstract way of writing that refuses, like Sbarbaro, to evoke context, in favor of exacting philosophical reflection on essence and language, and a baroque vitalism nurtured through contact with numerous cultural references from Greek and Roman antiquity to the modernity of the 1970s. This palimpsest-writing method is shared by other poets of his generation, united under the name of *Novísimos*. In the 1990s, his poetics, always aiming for intellectual emotion in the direct line of German Romanticism, broke down and took a new direction toward existential meditation and lyricism in the formal space provided by the elegy. It is within the framework of this new reflection on the self and the masks of identity that an allegorical landscape gradually takes shape, in which lichen becomes one of the recurring figures. Linked to a reflection on time and language, it is associated with memory ("Forms within the lichen/hours") and expresses what is permanent and resistant about existence ("example of resistance to destruction imposed over the course of time").[102] It is a force, the force of the elegy's perseverance: "*este dolor de líquenes*," this "grief of lichens" and these "lichens of darkness" that time leaves in us. Our face henceforth can no longer be represented by anything but a watercolor, water washing away and mourning an ever elusive identity:

> I see the leaves: they groan like rigging.
> Sails on the lichens of the sea.
> Nolde is deep within. In an image
> something always reappears: what is coming
> is all that I was.[103]

The expressionist aesthetic conveys this existential wound. The "lichens of the sea" are those from the beginning of time, or those red arborescent algae whose scientific name is *Chondrus crispus*. Lichen is associated with the liquid element, which highlights the wordplay reinforced by the Spanish "*líquenes/líquidos*" "liquid lichens."[104] "Brilliant lichen" belongs to this "childhood garden," to its "fleeting geography." In *Semáforos, semáforos* (1990), there's a painting by Gauguin: "the snow-covered watercolor/made by a bed of lichens."[105] Thus, by way of *Late Hymns* (1999) and *Steps in the Snow* (2004), an originary geography or landscape is constructed, not a surrealist one, but allegorical and intellectual, with recurring elements: seaweed, moss, lichen, the sea, mother-of-pearl, leaves, salt, gold, light, composing a kind of "vivid nature." They come to replace the table of conceptual oppositions from the first seven collections.[106] Perhaps there is a distant echo of Borges, and the echo of the unmade "I" who speaks in the poem:

> Echoes, undertows, sand, lichen, dreams.
> I am nothing but those images
> Shuffled by chance and named by tedium.
> From them, even though I am blind and broken,
> I must craft the incorruptible lines
> And (this is my duty) save myself.[107]

With Jaime Siles, lichen appears as the existential symbol of permanence: it is what resists in the face of time and the broken "I," what perseveres through its toughness, in the same way as mother-of-pearl, as concretion. Moreover, the words *liquen* and *nácar* ("mother-of-pearl") convey this toughness through their hard sounds: they are often used in the play of resonances (with *líquido*, *zinc*, *eco*, and so on). In this sense, lichen is an anti-conceit, an anti-baroque fantasy: it stands in opposition to the unsteady flame, the bubble, the hourglass, rust, ashes, the flower that withers, or the fruit that rots.

> Invisible voice
> of indivisible drops
> fire reaches you
> in the voice of water.
> Marble, moss, lichens
> that were
> your memory of long ago
> in water.
> In forms of fire
> which are light
> the figures of fire
> which are water.[108]

Lichen escapes death, "anti-destiny" fantasy, Malraux would say, a destiny very often linked to that of marble (the words resonate in Spanish) in Siles' poetry. Lichens, like marble, as well as moss, call up tombs, monuments of permanence. Such a landscape runs through Siles' writing. Having done research in Latin epigraphy, he worked on archeological ruins. But most importantly, his poetry itself is tomb, elegy, poem-cenotaph, and song of dissolution: that of the nothingness of being, of identity.[109] As Henry Gil has demonstrated, the poems in the collection *Música de agua*, in particular, often resemble calligrams or Latin *carmina figurata*, often taking the form of urns or bowls.[110] In this sense, the writing-epigraph develops and attaches itself like lichen.

The form of the poem also plays once again with a baroque aesthetic that cultivates plant metaphors, particularly to refer to writing. In the collection *Himnos tardíos*, Jaime Siles takes up a form typical of the Spanish baroque, the *silva*.[111] Created by Francisco de Quevedo, a *silva* is a series of lines of eleven and seven syllables, with no determined length and with consonantal rhymes or without rhymes. This structure allows the poet great formal freedom and comes close to free verse. Thus it is the most modern of the baroque poetic forms, and it has become the model for the long lyric poem in modern Latin American literature (Rubén Darío,

Octavio Paz, and so on). This freedom is contained in its name: *silva*, which means "forest" in Latin, and becomes *selva* in Spanish. The walking in Siles' baroque poems turns into wandering, among a forest of concepts.[112] *Silva* becomes *selva* (a wordplay that we find with various baroque poets, like Luis de Góngora in *Solitudes*, who often used this poetic form for pastoral subjects). This forest metaphor for poetry comes down to us from antiquity, moreover. Stacius composed *sylvas* in the first century, a series of poems unrelated to one another, a mix of verses. The form was taken up again in sixteenth-century France under the name of *boccages*. The poem turns into tree, darkness, its meaning wanders: solitude and elegy of the lyrical subject. "*Confus errando en bellas soledades*" ("Wandering, confused, in beautiful solitudes") wrote Espinosa Medrano (1630–1688). "Nature is a temple in which living pillars/ sometimes emit confused words," Baudelaire responds.

Nuno Júdice on the Lyre of Lichen

At the end of the twentieth century, from the western side of the Iberian peninsula, Nuno Júdice (born in 1949) linked lyric speech to lichen even more directly. In a collection published in 1985, he connected the two words by playing with their sounds (the initial "li" syllable they shared is accented in Portuguese): *Lira de Líquen* [Lyre of Lichen]. Thus he emphasized the close proximity of the organism that lives in hostile conditions and the lyric: lichen becomes the metaphor for poetic speech.

Just as with Siles, after writing in a style marked by textualism and abstract thought, Júdice reinscribed the lyric into everyday reality: "song in the thickness of time" (the title of a 1992 collection) or "meditation on ruins" (1994 collection). *Lyre of Lichen*, the collection that defines this reorientation, summarizes the idea in its title: the lyre, the traditional instrument of lyric and pastoral poetry since Apollo and Orpheus, ends up being composed of lichens, of usnea strings, or strings worn away by lichen, a ruined, abandoned lyre, crude and decaying, that time and nature have gradually overtaken – lyre from an antique statue of Orpheus pulled from the depths of the sea or discovered at an archeological site. Or else – and herein lies all the ambiguity of the title's "of" – the lyre plays a melody of lichens (a "music of cypress," Júdice writes): a ruined sublime. "*Lyre of Lichen*, emblematic title, ironic, paradoxical union of the sublime and the fragile."[113] In short, the lyre is inscribed in passing time, the time of memory and real life, it resonates with the history of civilization and poetry, and exhumes especially the genre of the elegy: the "lyre" as "song," "lichen," as the "thickness of time."

> [...] I search for the walls in the darkness,
> submerged by liquids up to my neck – water, mud,
> silt, lichen, target with a hyphen – and my head beats
> into the sinuous backs of fish, is scratched open,
> spurts green blood that glistens when it beads
> on my forehead. [...]
>
> "The Death of Kleist"[114]

Those images within the dashes "*água, lama, lodo, líquen*" ("water, mud, silt, lichen") evoke memory. They are part of a memory of childhood in Algarve, in southern Portugal, at the crossroads of personal and collective history. They attest to a time gone by, to the awareness of *tempus fugit*.

> These images of lichen, moss, and silt are linked to memories of childhood – my grandparents' house, very dark and moist, in which the walls of the inhabited parts were often like this after winter, and then the nearby waterways where I saw crabs and birds along the muddy banks.[115]

Although the word "lichen" itself rarely reappears in the collection, all the poems are inspired by this muse of the night, this hibernation, this anamnesis-katabasis, permeated with images of dead men and women, absences and waiting, twilight and graveyards, emptiness and the abyss.

> Angel, whose dark face
> appears to me veiled with tenderness,
> why do you rest your wing, so pure,
> in the wet, cold, hard stone?
> [...]
> Help me to hear, in the wind,
> the echo of an ancient lamentation:
> murmur left by the lover
> in the tomb that complains of the silence;
> [...]
>
> "On the Anonymous Lovers of Père Lachaise"[116]

Júdice's poetry thus traverses the night of memory and tries to describe it. The night of the dead, of the death of forgetting, from a moment of nostalgia: the gaze of a "late" poet. But the attempt is condemned. "Naked goddess, made of the matter of the dead,/ flesh eaten away by the slow lichen of the summits" (*Meditation on Ruins*). These images of decay occur frequently: ruins, leprosy, corruption, dissolution, rot.

> The remains of light, of dispelled vespers,
> at the horizon form bloody lines;
> clouds wander in my soul, worn out,
> over a face marked by pale dawns;
> the music of the Barbary organ, muted
> by the dull coin of the blond child;
> and the source of light, on the bananas – light woven
> from red reflections on the rotting foliage;
> [...]
>
> "Nocturne"[117]

Decadent music ("Barbary organ"), "muted." In this sense, lichen could be a metaphor for these "remains," "vestiges/of voice" and "traces of shadow," memories that rise to the surface, embryos of lives, vague and faltering images from the lyre strings.[118] It could represent language's inability to recapture the past, except in this fragmentary, ruined way. This nocturnal, crepuscular lichen is not, however, a force of resilience and life, but a trace of accumulated, static, melancholic time.

> At the end of autumn, one knows it's the end
> when the grass disappears under dead branches,
> when the sky takes on the color of rust and
> the wet fields smell
> like seaweed, what the tide brings in, rumors
> of the soul.
>
> "Viewpoint Indicator"[119]

As for Siles, lichen is one of the elements in a landscape both autobiographical and allegorical (and literary), along with water, silt, moss, seaweed, kelp (the *salsugem* of another Portuguese poet, Al Berto).[120] It speaks of origin, memory, time, but in this case, conversely, of their inconstancy and decadence. These images may evoke the baroque aesthetic, the Romantic paintings of Casper David Friedrich, who loved landscapes of ruins and dark forests, or the metaphysics of Dutch painter Albert Carel Willink, but they also attest to the exhilaration linked to the context and success of surrealism in Portugal, that marked all Portuguese poetry of the second half of the twentieth century, from Al Berto and Herberto Helder through to Nuno Júdice.

Antoine Emaz: To Hold, by Dint of

It is a body that stands there, painfully, in the stiffness of its angles, upright and silent, struggling and straining against the heaviness, as thin and gray as its metal heart. Giacometti's sculptures and drawings makes me think of the voice at work in the poems of Antoine Emaz (1955–2019). A body all skin and bones, spindly, crossed out, reduced to a few marks nervously penciled in. Emaz's poetry is also wire poetry, not wire fencing, but the experience of brusqueness, fatality, concentration, of struggling or twisting in the face of what happens. It seeks a "vertical dignity" beyond melancholy:[121]

> to live
> without much hope except
> to hold the no
> not to end
> head low[122]

Emaz's voice is always arising in opposition, facing or against something, finding expression only because it encounters an obstacle ("cannot *but*," "*without* much hope *except*," "to hold the *no*"). It is from this lyric negativity that a minimal, vegetative voice emerges. Demonstrative or presentative, it articulates a state of the body, a body that holds despite everything, like an imperative ("I must hold onto my books; they must hold," "one holds still/standing"), seeking verticality, like Sisyphus. In 1963, Jacques Dupin wrote of that same daily struggle in a prose text entitled "Lichens":

> To climb you, and having climbed you – when the light is no longer supported by words, when it totters and crashes down – climb you again. Another crest, another lode.[123]

A body that resists the monotony of the quotidian, the teacher's paperwork, winter and rain, the media and the economy, external solicitations, "work fatigue," a world inhospitable to poetry, but also and especially disease, old age, all that comes from within, from the body, the void, as much "external gunfire" as "internal cracks," in short, "this life – skin of sorrow imposed on the greatest number." The aperture of his poems lies exactly at the torn interstice (the "no," the "except," the "but"), the narrow space of strangulation, a call for air faced with the equivalent *anguish* or *angina* that threatens. Hence the frequent images of walls and silence, powerfully illustrating here a poetics with a literal, realist bias:

> This world is foul with stupidity, injustice, and violence; in my opinion, the poet should not respond with a salvo of dreams and a magic spell of language; there is no forgetting, fleeing, or distracting oneself. You have to stand with those who say nothing or are reduced to silence. So I write beginning from what remains alive in defeat and without a future. If it isn't easy to write without illusion, it would be harder to stop and suffer in silence. So [...] I like to think of poetry as lichen or ivy, with the small hope that the ivy will get the better of the wall.[124]

Hence this impression of "tight-lipped words": lines reduced to a spinal column, to bone, ellipses, syntactical ruptures, restrictions, suspensions, absent articles. The writing style is close to André du Bouchet's in this way, and paradoxically aims at providing air again, and meaning. Gascar mentions the affinities between lichens and scars.

The experience of resistance for Emaz is not a kind of "stoic heroism," or sadness, there is nothing of the epic or effusive lyricism here ("I am not going to lyricize this void," "things are as they are," "the rose is not lamenting [...] it is a rose"), but rather the expression of an acknowledgment, on the human level. A literal elegy, close-cropped (Perec says "to see flatly"), where the aesthetic follows mundanely from existence, without sadness or consolation, without music – a monochord lyre. I don't know for certain if Emaz read Sbarbaro (who spoke of "poor life," see above p. 120), but this assessment is expressed through a shared image:

> This is not an aesthetic of the poor, it is a fact: life is poor. But one would be wrong to think that the close-cropped is without depth. [...] To stick to the roughest grasp of what is. [...] I am what is, nothing else.[125]

Lichen becomes the symbol of this experience of taking hold. When Antoine Emaz tries to describe himself, to confront himself in the mirror, lichen is the organism he identifies with in his

Autobiographie en végétal (2009): "I don't know if I truly have the head of a lichen, surely not, but I envy its persistence." Following Hugo's eagle, Baudelaire's albatross, Corbière's toad, Roubaud's earthworm: lichen. It recurs frequently in Emaz's writings, both practical and theoretical. In the process of becoming a plant, as already perceived by Rousseau (let us recall his magnificent lines: "My head is filled with nothing but hay; I am going to become a plant myself one of these mornings"). Lichen for Emaz is valuable as a "word-program," a metapoetic allegory. But there is no brandished emblem here, ever; this is not the symbol or flower that triumphantly adorns the blazon of the medieval poet-knight, but rather a simple image to speak of the struggle of existence. No longer the beautiful rose, but the meager lichen. To say just that, without embellishment. Always the antihero, the poor, cursed, minor poet. Confronted with modern ugliness:

> My bond with nature [...], perhaps an avatar of Romanticism, certainly the *experience* of a difficult social nature and finding comfort in a relationship to the plant world. There is no spiritual background; only being alive among silent living beings.[126]

No spirituality, then. But the experience of silence and solitude through a privileged fusion with nature. We hear echoes here of Sbarbaro, Jean Follain, Guillevic, and Reverdy. But even if there's a claimed bond with nature, we are worlds away from Romanticism.

> The poet is a tree that may be blown about by all winds and watered by every fall of dew; and bears his works as his fruit, as the *fablier* of old bore his fables. Why attach one's self to a master, or graft one's self upon a model? It were better to be a bramble or a thistle, fed by the same earth as the cedar and the palm, than the fungus or the lichen of those noble trees. The bramble lives, the fungus vegetates. Moreover, however great the cedar and the palm may be, it is not with the sap one

sucks from them that one can become great one's self. A giant's parasite will be at best a dwarf.[127]

Thus, according to Hugo, the poet must be inspired directly by nature ("nature is the truth," a "foundation" whose words must be heard) and nothing else. The poet is a colossus and a free creator, not an epigone or a lichen (seen here as a parasite) that profits from models. Without "support," the poet lives intensely and vertically.

With Emaz, lichen also gives its name to two collections of notes: *Lichen, lichen* and *Lichen, encore (notes)*. The repetition of the word and the extension via adverb speak of this stubbornness to find expression, to persist (that "green stubbornness" to live, to quote Lacarrière). In this sense the "frayed thoughts" scattered throughout the notes, like aphorisms but poor, damaged, lacking transcendence ("thought ground up, like meat"), are so many branches of lichens that keep struggling, an experiment with pushing the line, which is trying its best, the repetition, the space won through exhaustion, to the very end of the line, of the breath: "at the end of the race, without having run." Thoughts that are formless, disheveled, "hirsute" (as Cuban poet José Marti would say),[128] like lichen: "somewhere between a bit of everything and anything at all: description, poem, journal, witticism, criticism, sketch. [...]" Close to Montaigne's essays in that way, tailor-made to express his "unformed and unresolved fantasies." The horizontality of these lichen, these frayed threads, are like the verticality of Giacometti's spindly lines, like the cursory marks of Hélène Durdilly and the ink drawings of Djamel Meskache that punctuate the notes, like the arborescences of Xenakis.

This "lichen voice" finds expression in Emaz's poems as well as in his thoughts on those poems. As with Sbarbaro, lichen articulates the frayed nature of thought and prose, the fragments of notes and beginnings of aphorisms ("drifting from head, scraps, islands"), but also the simplicity and poverty of the poem. "Poetry of not much, of what remains, resists, and must suffer: viaticum of words." "Lichen speech" is expressed through great economy of words, like the minimal, polar, extreme vegetation that inhabits hostile regions, breaking through the silent surface: poems as raw reports, what is broken off, ruined, what remains, what one manages to wrench free in the end. "Lichen speech" takes place over the long haul, cultivates slowness ("living slowly"), and lets itself go, giving up metric rigor: no lines, no "text," frayed bits.

> Inside operates in a much slower time than outside. You must learn to wait, to let emerge what must. It's especially important not to force things: allow for fallow periods.[129]

Humble and "non-lyrical," "lichen speech" conveys a pathetic vision of life, a "small life" (Pierre Michon). "Be low" writes Canadian poet Ken Babstock. As for other poets and artists, it is the image of what resists, persists, insists: a poetry that is "minimal and without glory, but resistant and persistent there where nothing grows." Perhaps for Emaz it is less about resistance, which can be too glorified or heroic (Brenda Hillman talks about the "baby epics" of lichens) – unless minimal. It is the "persistence" of lichen that strikes him – its "confidence," as Giono said. "There is no fervor/ just/ remaining" (*Peau*). More like stubbornness, endurance, inertia, attempt. "It is less a matter of optimal maintenance than of advancing," "I don't advance, I try to get by," Emaz says. Less a poetry of hope than of effort, of tension. "Need/ for this graveled, stony thickness, not easy, need/ perhaps for this resistance of poor ground" (*Lichen, lichen*): like lichen, Emaz's poetry is "extremophilic." "I don't hope, I try my best," André Frénaud wrote. "Lichen speech" seeks above all to be anchored, even minimally, even locally. To be anchored and provided air. No need for travel or epic to confront life and writing: "Basically, it's as crazy to go to the ends of the earth as it is to stay at home. The important thing is to venture, to try"; "Language is not necessarily Himalaya; it can be Beauce." Let us remember Rousseau's voice here: "To my mind, the greatest charm of Botany is to be able to study and know the natural world around oneself rather than in the Indies." The experience of solitude, inhabiting nature, rooted in the local, but without enthusiasm or epic fantasy, without imagination or dream. Finally, "lichen speech," like the organism itself, must rely on memory, on feeling.

Emaz's minimal poetry thus constitutes a kind of "night watch" ("the calm of a word like 'night light'"), of muteness (*Voix basse* [Low Voice] is the title of a 1995 collection). Emaz compares current poetry to lichen: it can be on standby, but it keeps watch over what surrounds it. Writing is "watching over language, [...] a minimal form of action and hope."[130] Meager, unobtrusive, isolated, it persists in hostile environments and sustains hope.

> There is *a form of minimal resistance* in the simple fact of still writing poetry, despite everything. For how many people? For one hundred, two hundred readers? It's crazy! One is no longer part of the logic of this world. And nevertheless, I continue to spend my life, my time, working for that. It is a real commitment. As to the hope for poetry, I have no fear on that count. Poetry has already been degraded enough in the economic sphere, the publishing world, any readership, that nothing more can happen to it. [...] Thus, hope – we're talking about reasonable hope – exists. Poetry is not like fireworks that light the sky, are seen from a distance, and quickly fade. I view it rather as a *night light* or lichen. [...] Lichen is used to hostile environments. When living conditions become impossible, it sinks into lethargy, to be reborn when they become acceptable. It seems to me that poetry goes through such phases, without ever disappearing completely.[131]

Olvido García Valdés: Green Glimmers

Influenced by the phenomenology of perception and Zen philosophy, the Spanish poet Olvido García Valdés (born in 1950) makes the poem a space of opening, listening, looking: a window on the world and, in particular, on nature. It testifies to an exercise and an ethics of attention, as well as an ecopoetic sensibility. One must "let things enter" in the poem, things from memory, from everyday life, which immediately necessitates a new relationship with time: "long" time, that of the pause (the *demorada*), of a present outside of time that contrasts with the currents of the modern world. "I believe that attention is the phenomenon of stopping, of letting things enter, arrive."[132] Attention is focused in particular on two special objects: nature and memory. Nature, with its preferred places like the garden and the park, and adopting a political stance aimed at rehabilitating the presence of animals ("Nothing better to do than look at the birds/ or look at the trees").[133] Memory, in the desire to explore the past, real life, its wounds, and so not to separate experience from the language of the poet receptive to the world and self ("my books are inseparable from me; each bears within it a part of my life.")[134]

It is in such a scenario that lichens regularly make an appearance. They show up particularly at the turn of the century in the collections *Caza nocturna* (1997) and *Del ojo al hueso* (2001). They are evidence of the power of life and adherence:

> [...] For the one who writes
> the long poem for the death
> of the father, conviction puts in order
> what language confuses and life
> annihilates. One's conviction is north,
> east, as if it were saying: in this case
> step back, look with more distance
> resonance
> at what is developing in the breast, lichens
> are born and adhere, tenacious, to the
> volcanic rock. [...][135]

Lichen represents volcanic speech, a "live" presence, a thread that goes back to the beginning, to memory: the nostalgia of being anchored.

> Speak of lichens, matter
> of memory, form
> that comes from the unformed
> through sediment and wear, trace
> and gesture of knowledge.[136]

But this image goes hand in hand with anxiety. These two collections are permeated with the presence of night and lurking death: in 1997, cancer struck the poet and brought with it surgery, chemotherapy, and radiation. Her collection, *Del ojo al hueso*, bears the marks of this: the light cast by lichen is very much that of the "mortal green" coming from night, a lost past, scanners, the dead bodies in baroque paintings, ghosts, and what persists ("the surviving matter – luminescent but pale, weak, often greenish – of ghosts," writes Didi-Huberman on the subject of fireflies.)[137] "The dream is sometimes only a spectral lichen," wrote René Char. "If/you were green, love, death/ you would be. With the thin/ and subterranean light" (*verde* [green] and *muerte* [death] echo in the Spanish). When García Valdés evokes lichens, her voice is fundamentally elegiac, ventriloquist or somnambulist: the struggle in the face of lost and decomposed words, "burned song" the Russian poet Anna Akhmatova would say.

> And I remember that at a certain moment I began to distance myself deliberately and consciously from my fascination for lichens, almost as if it were a matter of an objective manifestation, organic and vegetal, of the disease, organic and animal; a sort of external correlate to the malignant cellular proliferation that cancer involves.[138]

Lichen evokes the context of aphasia (during those years, the poet's cancer evolved simultaneously with a beloved's Alzheimer's: "mechanisms of the one who advances in the process of losing speech (and memory and the capacity for attention), even while retaining, in contrast, the threads of emotion."[139] Lichen as expression of the precariousness of language:

> The functioning of language is strange, so natural and apparently solid and certain, but so light, ethereal, devoid of roots, when something alters it (from simple psychological insecurity to neuronic and nervous short circuits). So similar to lichens (although, at the time, this parallelism that I've just laid out was purely intuitive).[140]

Jacques Lacarrière, Enthusiasm and Obsolescence

Jacques Lacarrière (1925–2005) acquired a taste for "back roads" at a very young age.[141] Fascinated by the Greek world and the East, he learned Hindi and traveled regularly in Greece, to Mount Athos in the 1950s. As traveler and writer, translating Herodotus and singing the praises of walking, like Rousseau, Thoreau, and Nicolas Bouvier, he declared in his *Chemin faisant, mille kilomètres à pied à travers la France aujourd'hui* (1974), that he wished "to reinstill in his readers the taste for plants and paths, the need for idling in the unforeseen, for finding one's roots again in the great message of horizons." Such a stance draws on an ecological awareness, a pastoral sensibility. Deploring "the loss of all real contact with the earth and living nature," he emphasized how "the Greeks and Romans had a primal sense that humans belonged to the earth, whereas today, we experience the opposite."

Lacarrière's poetic work is filled with both physical and metaphysical, natural and spiritual inspiration, as indicated by that gaze toward the horizon. Trying to look hard at and describe the mineral and vegetal world that surrounds him in order to celebrate its mystery, Lacarrière defines himself as a "rural poet" and thus stakes claim to a form of obsolescence. Which is also the obsolescence of poetry in a postmodern era in which discourse abounds on the disappearance of poetry, a certain kind of poetry:

> The poems that follow are a homage to the natural world from which we arise and whose language we have totally forgotten. All belong to the same family: bucolic, agrarian, forestial, telluric, aerial, nebular, cereal. These are poems out of season, poems of prayer. [...] Rural poems. They have sprung from my country life, my hours of walking, daydreaming, careful observation, weariness, terror or enthusiasm before the detail of every detail of nature. [...] Poems rising just above ground, sometimes. Or cut to ground level. Down to the smallest bit of the least clod. I want nothing of the minuscule to escape me. Nothing of the hidden or the occult. The truth is that I have never been able to resign myself to the definitive death of elves and nymphs. [...] Thus these poems might be stelas to their memory that I sense, of course, to be full of obsolete fervor."[142]

Lacarrière describes his "rural" poetry here in deliberately Rousseauist terms: proximity to nature, attention to detail, awareness of obsolescence. Let us recall the famous quote from *Reveries*, when Rousseau wanted to give its rightful place to "every grass in the meadows, every moss in the woods, every lichen covering the rocks – and I did not want to leave even one blade of grass or atom of vegetation without a full and detailed description." But it is coupled with inspiration and mystery. Its obsolescence lies in the form and the poet's taste for classical literature (as is true for other postmodern poets, like Jaime Siles (see above, p. 136), or Jude Stéfan, born in 1930, also influenced by antiquity).[143] There is also that attention to rural modesty shared by the lichen poets (Sbarbaro, Rousseau, Thoreau). There is wonder, often religious, in the face of the complexity and mystery of living creatures and smallness, which is also evident in the observations of certain artists and scientists like Thoreau or botanist Van Beneden.[144] Observing diatoms – micro-algae – Van Beneden responded with this lyrical outburst:

> The best microscopes do not always reveal all the delicacy of the designs that adorn these admirable organisms; even

the instruments of the best houses are hardly sufficient for bringing out the infinitesimal extravagances that decorate these Lilliputian carapaces.[145]

In 2013, Canadian poet Lorna Crozier expressed this same sentiment regarding lichen:

> Something that comes close to holy:
> you must fall on your knees
> to see it clearly, weather's hallelujah
> turned to Braille. [...] [146]

Poems and stelas (we may think here of another poet-traveler, Victor Segalen): Lacarrière likes to muse about trees, flowers, sand and various stones and crystals, dendrite, erythrite, fluorite, cobalt, making the words ring like an epic song of nature. This time, we may think of Roger Caillois and his *Stones*, and especially Aimé Césaire ("I toss about/amidst the tender milk of lights and lichens," he writes in "High Noon") whom Lacarrière admired and with whom he shared this taste for nature, for incantatory poetry that plays with the breath, orality, and words (wordplay of sounds, registers, etymologies, unusual words).

In the early 1980s, a small collection was published entitled *Lichens* (1983–1985). Lichen becomes the object of celebration and fanciful interpretation, deliberately allegorical and surrealist: lichen, "confession of the unconscious of trees."

> Horizontal agitation of words. Lichen. Casket of the clear morning of peat. Lichen. Hunger of the cervidae, anfractuosity of the cold. Lichen. Rose of the finger on the trunk, rosary of the granites. Lichen. The reason for the little that endures, the reason for the nothing that remains. Lichen. Spine of questions, thorn of sky. Lichen. Undergrowth in haying time, bush of unreason. Lichen. Expat from the time of the clear morning of peat. Lichen.
> "Lichen III" [147]

This prose sings its song in the form of a series of definitions/periphrases. The word "lichen" repeats like the "amen" of a prayer to creation. "Horizontal agitation of words," lichen as the line. In "Lichen IV," the linguistic hymn becomes specialized. The poet chants the vernacular names of various species of lichens, interspersed with slightly systematic periphrases:

> Glaucous cetraria. Veined peligera.
>   Orange calplaca. Alectoria
>   *Allegory of harsh seasons.*
>
> Umbilicaria pulposus. Fleshless usnea.
>   Lobaria pulmonaria. Cladonia.
>   *Caledonia of the dry springs.*
>
> "Lichen VI" [148]

"Lichen IV" praises *Graphis scripta* and lays out the poem-lichen metaphor, signs simultaneously for both nature and autobiography:

> Calligraphy of the steppes: tundras of the elegy,
> calendar of clouds, parchment of bushes.
> And on your foam, the runes of the sea!
>
> That you trace the length of the trees?[...]
> This cartage of signs in my life?
>
> "Lichen IV" [149]

From the Hebrew alphabet or Chinese character in Lamarck to Sbarbaro's cuneiform, metaphoric musings on the fascinating lirellae of *Graphis scripta* transport us here back to runes (see above, p. 56).

*

In the Western poetry of the last few decades, lichen thus speaks of an existential crisis specific to the postmodern era (the disillusion and nostalgia of the subject who can no longer find or define her/himself). It also embodies a poetry turned toward the real, even the prosaic, or a kind of pure immanence (let us recall Lacarrière's magnificent "poems rising just above ground," "cut to ground level," lichen as "horizontal agitation of words," or Emaz's "poor poems," "things are as they are"). The modesty of the poem, of the poet without a capital p. Lichen thus testifies to renewed attention, increased tenfold, to nature, which updates the genre of pastoral poetry by seeking to relocate language there. Lyric poetry at the turn of our century thus brings together existential, poetic, and ecological crises: it seeks to question and articulate the ethical need for a new relationship with the world.

The image of a poetry struggling to get by in hostile times, but persevering, resistant, and in a state of watchfulness, underground or minimal, has been powerfully present in discourse since the 1980s. The elitist withdrawal into a kind of conceptual poetry, and the socio-historical, editorial, and media-centered moment, discouraging to the traditional poetry audience, are often blamed.[150] Lichen: "holding on, by dint," Emaz would say. The positive, active force of poetry, its vitality despite everything, are often invoked by its defenders, from writers to readers, publishers, and reviewers. The review *Lichen*, created in March 2016 by editor Élisée Bec, insists precisely on that vital force inherent in poetic speech.

> It was while admiring the lava flows on the slopes of Piton de la Fournaise that I had the idea for the review's name. [...] To be able to grow on blocks of stone that were only tongues of molten ore, I think that demands respect!
> Élisée Bec, interview, 2016[151]

Here lichen becomes a symbol for regreening and renewal, *regain*, capable of inhabiting and repopulating the most hostile places, covering lava barely three or four years after it flows. "Post-volcanic" speech, then.[152] Or post-apocalyptic, one could say, which, in Bec's *Lichen* review, favors short, even minor forms.[153]

"Insurrection of the Humble"

Dormant for millennia, lichens are now located at the heart of artistic and scientific discourse that is trying to conceive of a new global articulation within the living world and a new attention to the environment. Becoming "subjects," in books as well as in sentences, appearing as new "beacons," these incredible forces of life and resistance are taking on the power of subversion.

Because of pollution, lichen continues to abandon cities; in 1975, three years after Gascar's *Présage* appeared, Pasolini made the same observation regarding the insects called *lampyris* (fireflies and glowworms).[154] We still saw them in the gardens of my childhood on summer evenings, flitting about the boxwood bushes or on the lawn. Georges Didi-Huberman begins with this idea in his essay *Survival of the Fireflies* (2009), written as a tribute to Pasolini. Ecopoetics are inseparable from politics in it. Didi-Huberman analyzes the contemporary loss of a basic principle of pleasure and humanity. The minimal light of the firefly embodies a force of resistance and affirmation, a "popular spirit" in opposition to the saturated, flashy, spectacular lights of the media industry (this is Bernard Noël's "*sensure*," blinding us to the meaning of words), to the "consumerist dictatorship" of desire. For Pasolini, fireflies were "human signals of the innocence annihilated by the night [...] of triumphant fascism."[155] Emphasizing "the particular vitality of so-called periods of decline," Didi-Huberman stresses that it is the darkness of the world itself that makes it so these faint glimmers are not outshone and become visible, spread their light: "the living dance of the fireflies occurs exactly at the heart of the darkness." He thus formulates this moral imperative:

> The fireflies: it's up to us not to see them disappear. We, ourselves, must assume their freedom of movement, the retreat

without withdrawal, the diagonal force, the ability to make particles of humanity appear, to make the indestructible desire appear. We must, ourselves [...] become fireflies and thus form again a community of desire, a community of flashes shining out, of dances in spite of all, of thoughts to transmit. To say *yes* to the night all crossed with glimmers and flashes, and not be content merely to describe the *no* of the light that blinds us.[156]

*

The "night all crossed with glimmers" is also the one explored by artist Yves Chaudouët. Beginning in 1997, he took an interest in deep-sea fish. It is also the marginality, small size, and resistance of these fish, sometimes living at depths of more than ten thousand meters, that fascinates the artist, and in particular the lights of lanternfish, as their function remains mysterious. Let us recall Pierre Gascar's image comparing lichens to deep-sea slugs, the *Oneirophantes*, those great lickers of the ocean depths.

Why this interest in submarine, subterranean life forms, in those subversive lives that, because they are condemned to the darkness, resist and develop dialectically, creating their own luminous power? These are also the terms that writer Marcel Schwob used to talk about slang in the late nineteenth century:

This language has been decomposed and recomposed like a chemical substance; but it is not inanimate like salts or metals. It is compelled to live under special laws, and the phenomena that we notice in it are the result of that constraint. Animals living at great ocean depths [...] lack eyes, but pigmentary spots develop on their bodies that are phosphorescent. The same is true for slang in the lowest depths where it moves.[157] It has lost certain language faculties and has developed other ones to take their place. Deprived of daylight, and under the influence of the environment that oppresses it, it has produced a phosphorescence by the light of which it lives and reproduces.[158] Synonymical derivation.[159]

A minor phenomenon of bioluminescence, a minimal form of resistance: like glimmers in the dark. Lichen as a *popular* language to cultivate. German critic Peter Stephan Jungk mentions the political dimension in the landscape work of Chaudouët: to show the minor, the abandoned, in images that are so many will-o'-the-wisps (here we return to Sbarbaro's image), *ignis fatuus*, those "foolish fires," erratic but alive, that appear as subversive bursts:

> His pencils, pens, brushes, his plates and inks, but also the camera and video camera that serve him as microscope and telescope, as vehicle to penetrate the micro- and macrocosms. In a world that has made its ideal the smooth success, the incontestable perfection, the irreversible, Chaudouët is fascinated by the small resistances in the form of hooks, contradictions, contentions.[160]

"Hooks" that are "hooked into" reality, which also evokes the rhizines of lichens. That is because, beyond the figurative, Chaudouët's aesthetic is never separate from an ethics, an ecological and political dimension. Which explains his empathy for certain natural elements: he repeats that he feels a "familiarity," a "sympathy" for lichens, he speaks of fish as "models," and praises their "exemplary nature."

*

Self-portrait of man as firefly, as lichen, minimal and symbiotic place of photosynthesis:

> let lichen kind be a leading light
> how as ever each oxygenation event
> obliges us with this solar commune [...]
> > Drew Milne "Welcome to the Biotariat"
> > [*In Darkest Capital*, Carcanet 2017]

Following the triumphant Enlightenment, we are living in an age of glimmers and night lights. Lichens, fireflies, sparks, "furtive fervors" (Marie-Claire Bancquart), will-o'-the-wisps, phosphorescences, lanterns of deep-sea fish: sources of tiny, subterranean vitality that are active precisely at dusk (collapsology), "fungi at the world's end" incarnating the principle of minimal resistance. This idea of resistance is found in Michel Butor's very beautiful quote, expressed not as light but as breath: "there are lichens representing *breaths* [...] despite the parades, tortures, poisons, famines."[161]

The age of weeds, of "neglected biodiversity" is thus proclaimed, for thinking differently about the globalized free world. As poet Georges Duhamel wrote, it may be a matter henceforth of giving voice to the "disdained plants," to the lichen of "dissidence," to "the insurrection of the humble"!

> Men fought among themselves and that discord lasted a long time. Then, mosses, mushrooms, lichens, parasites of all classes and all dissidences, all the disdained plants, all the humiliated plants, all the living beings without flags and without laws invaded the gardens and made it known immediately that their time had come. It had not yet come, but it would come in the end. The next time, perhaps.[162]

We can find this image again in Zola, in the guilty hallucination of Abbé Mouret. The "gnawing of the little ones" that he depicts appears as a destructive force, revolutionary and satanic, tearing away at Catholicism and its prohibitions, and ending with the total downfall of the religious edifice:

> A mighty shout hailed the downfall of the block of rock. Yet the church stood there firmly, in spite of the injuries it had received. [...] Then Abbé Mouret beheld the rude plants of the plateau, the dreadful-looking growths that had become hard as iron amidst the arid rocks, that were knotted like snakes and bossy muscles. The rust-hued lichens gnawed away at the rough plasterwork like fiery leprosy. Then the thyme-plants thrust

their roots between the bricks like so many iron wedges. [...] It was a victorious revolt, it was revolutionary nature constructing barriers out of the overturned altars, and wrecking the church which had for centuries cast too deep a shadow over it. The other combatants had fallen back, and let the plants, the thyme and the lavender and the lichens complete the overthrow of the building with their ceaseless little blows, their constant gnawing, which proved more destructive than the heavier onslaught of the stronger assailants. Then, suddenly, the end came.[163]

In Brazil, US poet Elizabeth Bishop (1911–1979) compared the precarious populations of *favelas* to the powers of resistance and resilience of lichens:

> On the fair green hills of Rio
> There grows a fearful stain:
> The poor who come to Rio
> And can't go home again.
>
> On the hills a million people,
> A million sparrows, nest,
> Like a confused migration
> That's had to light and rest,
>
> Building its nests, or houses,
> Out of nothing at all, or air.
> You'd think that a breath would end them,
> They perch so lightly there.
>
> But they cling and spread like lichen [...]
>   "The Burglar of Babylon"[164]

Finally, in the experimental collection, *Extra Hidden Life, Among the Days* (2018), in which she dedicates twenty-four poems to various species of lichens and accompanies them with photographs, US poet Brenda Hillman (born in 1951) compares lichen to the plight of workers, minimal and minimalist forms of life, jeopardized or in jeopardy, that resist and work in silence.[165] Here are the heroic figures of Rosa Parks and Rosa Luxemburg ("lichen reads the stone, as Rosa rides the bus [...]"). True gold is no longer that of Wall Street but that of a small wall covered with *Xanthomendoza* and its vivid orange thallus, while *Candelariella* resembles "burned/ -out and undocumented huts from/ an invasion of yellow [...]"

*

This surge of working-class resistance and vitality is part of the desire to conceive of an alternative society, intimately bound to the world to which we belong.

> But we haven't time, in this world of ours, to love things and see them at close range, in the plenitude of their smallness. Only once in my life I saw a young lichen come into being and spread out on a wall. What youth and vigor to honor the surface!
>
> Gaston Bachelard[166]

Lichen invites us to leave the beaten paths, to cultivate slowness and patience to counter a consumerist ideal based on desire and the acceleration of time, to adopt a model of minimal growth to counter the hubris of exponential growth at any price ("Shall we always study to obtain more of these things and not sometimes be content with less?" wrote Thoreau in *Walden*, foreshadowing his praise for lichen as model for an ascetic life), to rediscover a point of connection with the local to counter the intoxication of the frantic globalization that permeates each of our smallest actions, our vibrating phones and our social networks.

"The insurrection of the humble." To rediscover humility, not in the face of an inordinately grand, transcendent, and overwhelming nature, but to better link our actions ethically to our environment, our earth, to a "we." Humility is also, etymologically, that which brings us back to earth, to "ground level." Humility is human; humility is *humus*, our origin and our destiny.

Not by globalizing, but by *mondifier* (to hijack Bachelard's term). That is, not to pursue the *global*, what can affect the world as a whole, risking dissipation and uniformization, but the *earthly*, the depths, the complexity, and the ecology in the small things that surround us (which oppose precisely the idea of globalization and growth).

*

Since the end of the twentieth century, a discourse has once again developed that extols decentered and ruderal spaces: fallow ground, wastelands, abandoned areas, ruins, no man's land. This trend is not only the residue of Romanticism or "fin-de-siècle" decadence, but an active site that is trying to think about new models for societies and a mode of life better linked to other living beings. Following more or less radical ideologies. Boho-lichen?

Since the end of the twentieth century, an imaginative space has in fact developed, in particular in French poetry and echoing a certain modern poetic tradition (the decadent poets, Huysmans, Apollinaire, André Gide's *Paludes* in 1920, which claims the swamp for its setting, T.S. Eliot's *The Waste Land* in 1922), that I would, again, bring together under the name of *ruderal*: as if henceforth the poetic voice willingly inhabited a marginal, abandoned space.[167] Two tendencies in contemporary writing intersect here: a renewed interest in nature and the imagery of decadence (an aesthetic of "poorness").

In 2000, poet and critic Jean-Michel Maulpoix wrote that "lyricism is a wasteland: an undefined space without boundary markers where all sorts of strange things end up: scraps of the world or the heart, with no set value or significance. A wild place, disturbing and nevertheless familiar, where, at the opposite extreme from the museum or church, the most elemental community is reconstituted."[168] But this imaginary ruderal space asks just as directly for prose, for the marginal note, somewhere between fragmentary account or cartographic doodle. Let us think, for example, of Philippe Vasset's *Livre blanc*, which explores the white areas appearing on maps of the Paris region. Lichen, model of a written landscape, ruderal and fragmentary, on the order of the note:

> In exploring my wastelands [...], I had the secret hope that my disorganized and contradictory notes would end up in a text that resembles this ground turned over a thousand times and mixed with debris, those cobwebs that catch on ears and hair, and those fruits that grow without being watered or tended. [...] I hoped that, despite everything, something was writing itself, clinging like lichen to those poor, friable surfaces, growing slowly, without plan or message.[169]

Does our salvation lie in fallow land? Ruderal space must no longer be seen as a place separate from us humans, that we neglect and that escapes us, but as a place of connection. It allows us to think about social and ecological alternatives by creating empty spaces rich in potential. The ruderal space is also a place of resistance in that it shows how living species – native or self-propagating plants (coming from foreign environments and growing without having been planted) – adapt to human decisions, or manage, despite everything, to persist and invent new forms and new ecosystems. Paradoxically, these spaces, often polluted, disturbed, poor in resources (soil quality, light, water), and small in scale, are among the richest in biodiversity (as opposed to the "green deserts" that monocultures create in the countryside). Marc Jeanson and Charlotte Fauve stress this:

> The plants always win. In cities, the environments that might be the most hostile to them, they are ubiquitous, more surprising, more enterprising than elsewhere. Without even looking hard, you couldn't miss that green fleece edging the sidewalk. Within ten meters, it's not unusual to find more than twenty species

perfectly comfortable with asphalt. [...] The sidewalks of our large metropolises shelter a *melting-pot*; this is the place par excellence for the intermingling of flora. A common dandelion appears alongside an *Galinsoga parviflora* escaped [...] from South America.[170]

It is estimated that seven hundred and sixty plant species grow of their own accord in the city of Paris. Published in 2017 by Xavier Barral Editions, the invaluable *Flore des friches urbaines du nord de la France et des régions voisines* by Audrey Muratet, Myr Muratet, and Marie Pellaton, though not mentioning lichens, describes these species in this way:

> Wasteland and fallow ground do not constitute one specific environment, but rather an ensemble of multiple habitats. [...] They are spaces for the free expression of nature common to cities and their surroundings. The diversity of plants and animals as well as their fertile interactions make these sites refuges for an exuberant nature that does not let itself be managed or cultivated.[171]

French gardener, landscape architect, and botanist Gilles Clément (born in 1943) has given fallow land much thought in relation to defining his own practices, especially his "moving garden" from the 1970s (which he initially called "tamed fallow land"). It is a matter of unmaintained land, rapidly colonized by plant and fungal species that can develop there freely. The space is seen as a place of energies ("growth, struggle, displacement, exchange") that the gardener, whose first task is observation, simply tries to balance, cooperating with nature to direct them toward their best use and increase biodiversity: to channel the competition and promote mutual support "to do the most possible *with* and the least possible *against*." This pursuit of a sustainable landscape is conveyed especially through reliance on perennial plants, a policy involving bulbs and tubers. It was inspired in particular by the concept of the "wild garden" proposed by English gardener William Robinson in 1870.

(Could the "moving garden" be a possible model for the writing of the *essay*? Critic Jean Starobinski has spoken of "moving writing" with regard to Montaigne's *Essais*, interestingly. Couldn't the essay be this "natural garden," this kind of writing that is constructed as it is being written, this page and this fallow land that let ideas happen and thinking develop freely, with more concern for raising questions than for affirmation?)

In 2003, Gilles Clément developed the radical concept of "third landscape," to refer to "all the spaces as a whole where humans abandon the evolution of the landscape to nature alone."[172] It can involve "fallow lands, marshes, moors, peat bogs, as well as roadsides, river banks, railway embankments, and so on": wherever their evolution is left to chance. The idea is to highlight spaces and plant life that are usually overlooked and that constitute an important reservoir of natural riches, to observe their inner logic and interactions. Gilles Clément's claim for the landscape adopts political rhetoric:

> The term "third landscape" does not refer to the third world but to the third estate. It goes back to the words of Abbé Sieyès: "What is the Third Estate? – Everything – What role has it played to date? – None – What does it aspire to become? – Something.[173]

Lichens and ruderal plants are very obviously part of this. The Matisse Park in Lille (the central embankment is called "Derborence Island") experiments with this pursuit.

I'm also thinking of the "pocket forests" (*florestas de bolso*) of Brazilian botanist and landscapist Ricardo Cardim (born in 1978), who uses native flora to develop replantation projects, in the interstices temporarily forgotten by São Paulo's urban management, in order to reconnect the population with its botanical heritage.[174] These actions, which echo, for example, Alan

Sonfist's land art in New York in the 1960s (*Time Landscape*), stem from the fact that in the largest Brazilian megalopolis, no less than ninety percent of the trees present today are non-native species, imported by the wind or by humans for gardens, and most of them are invasive. The idea here is not to fall into some form of nativism or botanical nostalgia. In his *Éloge des vagabondes*, moreover, Gilles Clément strongly opposed such excess, such "a blindly conservative attitude," and advocates for "the multiplicity of encounters and the diversity of beings as so many riches added to the territory."[175] It is much more a matter of reinscribing urban landscape design into the local, into the context of Brazilian deforestation, of maintaining the balance and diversity of living beings. By beginning with archives, often private ones, of the city's inhabitants, and archeological studies of soils that let him rediscover, under the concrete, the primitive humus and seeds of species that once grew in these spots, Ricardo Cardim is showing how he can achieve incredible results: in two or three years, there are exuberant bits of forest springing up between the skyscrapers.

Micro-habitats

"The style of the flâneur who goes botanizing on the asphalt."
Walter Benjamin, *Charles Baudelaire: A Lyric*

Through its great power of resistance, lichen is one of the most widespread organisms, especially in our modern cities. Adapting to harsh conditions and disturbed ecosystems, it is one of the key figures in the micro-landscapes that we can observe in urban environments, even the most extreme ones (the smooth, exposed surfaces of traffic signs or meters, plastic tarps on the ground in direct sunlight, and so on).

In this way, it invites us to shift our gaze, to change scale. Botany speaks of "micro-habitats" to evoke these small living spaces: tree trunks, stones. [...] In the city, these interstitial micro-habitats offer a contrast to the dimensions of wastelands: these are corners or edges of buildings, old walls, cracks in the concrete and sidewalks, trees, roofs. They constitute a reservoir of life and biodiversity, and they subsist, assert themselves, despite and thanks to cities. These ecosystems have their "sociology," and such cohabitation can be symbiotic; each species has its preferred living conditions, its neighborhood of other specific species. The sociology of lichens is a branch of lichenology that focuses on studying lichenic behavior: if, and why, certain species live close together. This concept of micro-habitats is illuminating because it demonstrates that the city is not opposed to the idea of nature (or rather to nonhuman living species), the city is itself nature, a place of interactions, and other living beings are still there, growing, emerging through the cracks, sleeping underground, adapting to anthropogenic conditions, and only waiting to be appreciated. That's what landscape designer Ricardo Cardim is trying to do by reviving the Atlantic forest in São Paulo: to show the dormant power of plants within the very heart of the megapolis, to strip back the asphalt so that saplings can grow.

> All of our architectural structures are exposed to the forces of nature. It is only a matter of time before they are overgrown like ruins by moss and various fungi that can live for thousands of years. It is precisely these often neglected organisms that have an important ecological role to play.
>
> Oscar Furbacken[176]

With growing urbanization and globalization, these micro-habitats have become new living sites, pioneer spaces. Constituting part of the ruderal, they are places of resistance, in that the organisms growing there must adapt to particularly difficult conditions: air and soil pollution, great variability in moisture and light, reduced space, and so on. The adaptation strategies of these microcosms have great educational value.

On this scale, dripping water pipes, city lights, and various sunshades create unique climatic conditions. Consequently, within a radius of just one meter, we can encounter several radically different micro-habitats.

> Lichens, often overlooked, have the ability to survive extreme conditions thanks to their advanced symbiotic makeup. Species sensitive to air quality can be used to measure pollution over time. Some fungi are used for mycoremediation, a process that allows soils containing toxic waste to regenerate. Mosses and other small plants are frequent colonizers of any new territory, including human-made structures.[177]

As Oscar Furbacken proposes with his project *Urban Micro-habitats*, there is also the question of reappropriating the city as a space that mixes human constructions with spontaneous habitats of other living species, and seeing it differently: ethically, ecologically, in a way that seeks to recreate a link among living entities. As a new version of Walter Benjamin's "flâneur," the Swedish artist wanders around big cities, in particular the more "touristy" ones, among visitors and inhabitants, to pause before a microcosm that captures his attention, whether lichens or chasmophytic plants (those that live in cracks). He then sets up his macro camera to make studies of the motif, a very small motif, and thus very close to the motif: paintings of "micro-landscapes" that he calls *Micro-habitats* (see Ill. 10) and *Close Studies*. For ceramist Leo Battistelli, it's a matter of more closely linking nature and habitat, even if it means making lichen grow in living spaces, in keeping with Austrian architect Friedensreich Hundertwasser (1928–2000) and his "Mouldiness Manifesto":

> When rust sets in on a razor blade, when a wall starts to get mouldy, when moss grows in a corner of a room, rounding its geometric angles, we should be glad because, together with the microbes and fungi, life is moving into the house. [...][178]

*

Modern lichens. They were there at the heart of aesthetic and scientific preoccupations in the nineteenth century at the time when large cities appeared, with their coal emissions, their architectural reconfigurations, and the Romantic and mystical start of a return to nature that they prompted. For certain writers in particular (Thoreau, Sbarbaro, Gascar, Emaz), lichens are the mirrors of existential quests tied to modernity and the postmodern age: they reflect metaphysical as well as physical anxieties, provide an image of the poet's condition, between melancholy and resistance, and offer an ethical and political model. They speak of the current state of poetry, marginal and marginalized, on standby, but surviving, resisting.

They attest as well to the emergence and evolution of an ecopoetic discourse, in literature and in the arts in general, that is still in dialogue today with a Romantic legacy and finds inspiration in Zen thought, even as it seeks to renew environmental receptivity. In Western art, a renewed dialogue with *physis* is taking shape, a neo-Romantic concept of nature but sometimes stripped of the transcendent and the sublime, and replacing, in any case, the modernist romanticism of technology and progress.

This new attention to living beings and to lichen calls into question our industrialized, urban world, even our post-apocalyptic state. It considers the need for a return to the earth, the small, the local. It reflects on those places of resistance and decentering (of changing scale as well) that constitute ruderal spaces and micro-habitats, that can allow for the awareness and the founding of a new ecology that would link the human and the city to all life, as well as a new ethics and politics. To make the ruderal a center, to inhabit fallow ground: to think of humans within nature, to change the model of growth, blur the boundaries, multiply the centers, even within the smallest crack.

Figure 11a © Oscar Furbacken, *Degenerational Crown*, 2020, sculpture in polystyrene, white concrete, iron, and fiberglass, originally in color, Norrtälje, Sweden.

Figure 11b © Oscar Furbacken, *Beyond Breath*, 2012, series of cut-out photographs (with the painting *Rising* in the background), originally in color, Katarina Church in Stockholm, Sweden.

Figure 11c © Oscar Furbacken, *Micro-habitat of Rome 2 (Ageless Glow)*, 2020, inkjet print, originally in color, 52 x 80 cm.

Part 4
TOWARD A SYMBIOTIC WAY OF THOUGHT

> "Rights of symbiosis are defined by reciprocity."
> Michel Serres, *The Natural Contract*, 1992

Lichen's incredible resistance is due to a very particular biological phenomenon: symbiosis. During our first encounter, when, with the energy and innocence a new research project prompts, I entered Philippe Clerc's office where the species for his latest study lay drying, he said to me, "Do you know that we have just discovered, in 2016, a third organism within lichen?" How is that possible, a *third organism* within lichen? I picked up a sample of *Xanthoria parietina* and examined its dried yellow plate, with its apothecia in the form of tiny suction cups. Where could a whole little world be hidden? You need good eyes: lichen's dual nature was only discovered in the 1860s.

*

One of lichen's most spectacular characteristics is undoubtedly its symbiotic nature. It actually unites many partners into a single individual: a "photobiont" (the algae, ensuring photosynthesis and, for that reason, often located in the superior layer of the thallus, close to the light, and providing nitrogenous and carbon-emitting materials, sugars and proteins, to the fungus – which can also be a cyanobacteria) and a "mycobiont" (the fungus, which provides structure and thus protection for the algae, as well as water, carbon dioxide, and mineral salts necessary for photosynthesis). Sometimes a second "mycobiont" is added (a basidiomycete "yeast") that takes part in the synthesis of the famous lichenic substances (it was discovered because two species of the *Bryoria* genus were different colors but their symbiotic partners were identical), and other microscopic partners.

"Lichenization" is the invention of a porous lifestyle, open to cooperation. Lichen is considered to be the result of a nutritional "strategy" of the fungus. In fact, the fungus is a *heterotrophic* organism: incapable (like human beings) of making the organic material it needs to nourish itself. Thus it needs other living beings: it can make use of the decomposition of soil ("saprophytism"), parasitize other species, or combine with them, as in the case of mychorrhizea and lichens (as for algae, they are *autotrophic*: they produce their own organic material from inorganic material and minerals through photosynthesis). In this sense, lichen is a fungus that, in order not to depend on decomposition, cultivates algae, as though in a greenhouse, allowing it to live at greater heights. And it is because of this heterotrophy, which distinguishes it from plants, that the immobile fungus is essentially symbiotic: obliged to cohabit.

The photosynthetic activity of chlorophyllic plants is coveted by other heterotrophic organisms. Other "photo-symbiotic" groupings were later discovered, notably in aquatic environments: the Roscoff worm (*Symsagittifera roscoffensis*) and the coral polyp both cohabit with single-celled algae for their respiration and nutrition (in the first case, the *Tetraselmis convolutae* alga, ingested without being digested by the Roscoff worm and surviving under its epidermis; in the second case, the zooxanthella, of the *Symbiodinium* genus). In these cases, they unite animal and vegetable kingdoms.

*

In *lichen* there is *link*. For many years now, certain biologists, philosophers, and artists have seized on this organism to question the notion of biological individuality (*indiv*: in Latin, that which cannot be divided, like Leibniz's monad, Lucretius' atom). The idea of the oneness and sovereignty of the individual, which sprang from Kantian and Romantic ideas on the subject, was gradually undermined on the philosophical level by Nietzsche, on the psychological level by Freud, and on the linguistic level over the course of the twentieth century.

Once again, lichen is in full dialogue with modernity. From a biological perspective, symbiosis is a notion that was discovered and defined at the end of the nineteenth century, beginning precisely with lichen.[1] Since then, symbiosis has been observed among a great number of living beings, including humans, and on different scales (from cellular to ecosystems). Almost all beings live symbiotically, that is, in an interdependent relationship (mutual or parasitic) with other beings in their environments. This phenomenon was only recently considered in the history of the sciences. As biologist Marc-André Selosse notes, "long taught as a series of biological anecdotes, [...] symbiosis is not anecdotal. The scientific community was slow to become aware of it and it wasn't until the 1970s that conferences were held on this subject."[2]

Every *bios* is *symbios*. Every organism appears as an ecosystem in interaction with a "symbiotic retinue" (with which it cooperates or struggles), laying out the principle of a dynamic ontology.

> No organism lives alone, and each possesses a symbiotic retinue without which neither its physiology nor its ecological success can be understood. This retinue is practically always present, since without it, the organism dies or sees its competition reduced.[3]

The Politics of Lichen: At the Origins of Symbiosis

From a political perspective, lichen could be rehabilitated through discourse idealizing the concept of symbiosis by reducing it to a "mutualist" definition: not only do the partners, fungi and algae, live together, but they mutually benefit one another. That way of life serves here as a projection of a model of social harmony, essentially Marxist, perhaps aimed at establishing a "biotariat" (Stephen Collis).[4] Old images, coming from ideas on universal harmony, could thus be readopted, like this passage from the poet John Donne (1572–1631) that shows up in blogs and features Man in harmony with the All: "No man is an island, entire of itself; every man/ is a piece of the continent, a part of the main." Insularity versus globalism.

That is the case, for example, with Scottish poet and academic Drew Milne (born in 1964). His *Lichens for Marxists* (2017), published on line, brings together "poem-lichens": poem-emblems consisting of photographs of various lichens on which are superimposed slogans or lines of poetry that advocate ecological resistance to capitalism. These aphorisms, or rather these placards for poetic demonstrations, are in conversation with the history of modernity and the literary avant-garde, who have deliberately played with the codes of advertising materials and political tracts since the late nineteenth century. Let us think, for example, of José Asunción Silva, Blaise Cendrars, Guillaume Apollinaire ("You read the prospectuses the catalogues the billboards that sing aloud/That's the poetry this morning and for prose there are the newspapers"), Dada, the visual and sound poets. And just like lichen (often, in any case), the placard appears on a vertical support, a wall. The lack of punctuation clearly shows this rejection of closure, and the "we" becomes the new pronoun for the symbiotic subject:

> we the symbiotic alliance of lichen/ hold the evident truth to the self/ namely that all lives are not made/ the same and the carbon liberation/ front will be the death of all but/ the persistent solidarity of algae.[5]

*

But let us return to the origins of the word "symbiosis" and the inner "solidarity" of lichen.

The word appeared relatively recently. In 1825, German botanist Karl Friedrich Wilhelm Wallroth (1792–1857) observed entities present in lichens that he called "gonidia," and that were actually algae.[6] The dual nature of lichen was discovered, and thought of as such, for the first time in 1866 by another German botanist Heinrich Anton de Bary (1831–1888), and in 1867, by Swiss botanist Simon Schwendener (1829–1919). Microscopic observation of the organism's growth revealed that fungi filaments and algae nuclei developed at the same time. From then on, lichen no longer appeared as an autonomous kingdom, but as the union of two organisms: a complex that contained algae surrounded by fungi filaments. That is what Schwendener called the "algo-lichen hypothesis." At the same time, in 1867, Russian botanists Andreï Sergueïevitch Famintsyne (1835–1918) and Josep Wasilijevitsch Baranetzky (1843–1905) made similar observations by succeeding in isolating gonidia (algae reproductive cells in lichen) from *Xanthoria parietina* and *Pseudevernia furfuracea* and making them grow outside of the lichens. Until the 1890s, German and Russian scientists especially represented the leading edge of research on symbiosis.

In 1868, Schwendener wrote: "I believe that lichens are not autonomous plants but fungi (ascomycetes) for which algae, about whose independence I have no doubt, serve as foster plants."[7] This union was immediately viewed by Schwendener as an asymmetrical relationship: the fungus was a parasite of the algae that it held in the claws of its hyphae. A lichen is "a community composed of a master fungus and a colony of algae slaves that the fungi holds in perpetual captivity so that they provide it with food," he also wrote. In this period, scientific theories of associations were conceived beginning from animals and essentially took the form of parasitism, until new interactions could be imagined (commensalism, mutual aid).

This way of thinking about living beings was revolutionary. Lichens would no longer be "elementary" or "rudimentary" plants, but complex and particularly evolved structures. Plural individuals could exist. Thus intense controversy developed within the scientific community, between those adhering to the autonomy of lichens and the "Schwendenerians" who promoted the "algo-lichen hypothesis." Insults flew back and forth and publications became manifestos and pamphlets. There was also a generational divide. That was the case with Famintsyne, and again with Finnish botanist William Nylander who, until his death in 1899, refused to believe in the dual nature of lichens and decided to stop frequenting the French National Museum of Natural History in Paris where he did his work, arguing that the researchers there had taken up the "Schwendenerian" cause!

Thus 1867 was an important year. It was also the year of Baudelaire's death and the publication of the first book of *Das Kapital* by Karl Marx, who was trying to "decapitate" a vertical social body.

In fact, for scientists convinced that lichen was no longer an autonomous realm, the only subject for debate now was the nature of the relationship between the fungus and algae. Parasitism, said

Figure 12 © Nathalie Ravier, "Glossary of terms currently used in lichenology," prints on tracing paper, p. 19.

Schwendener. It would be ten years before this cohabitation could be conceived differently. The work of Belgian zoologist Pierre-Joseph Van Beneden (1809–1894) headed in that direction. In 1875, he spoke of "mutualism" in his treatise, *Les Commensaux et les parasites dans le règne animal*. Albert Bernhard Frank (1839–1900), curator for the University of Leipzig herbarium, coined the word "symbiotismus" in 1877:

> The relationship in question is something much more than simple parasitism in the usual sense, because while we imagine that originally or generally lichens lack gonidea, in fact, the parasites and the host are united from the start to constitute a new unified organism. [...] From the union of the two organisms ... the formation of a specific new form results. [...] In any case, this is based solely on "living together" and that it why we can recommend using the term *symbiotismus* to refer to these cases. The phenomenon is not to be considered completely parallel to what happens with parasites of animals, like certain parasitic fungi and in particular those that make galls. [...] *A relationship in which the parasite also cares about the nutrition of its host takes on a different significance from parasitism.*[8]

Parasitism was an outdated notion because it did not allow for the formation of an organism beginning from an initial union of two different organisms, nor for the idea that the host organism (the algae) could benefit from this arrangement. The following year, Alsacian German botanist Anton de Bary defined "symbiosis" as "the shared life of organisms with different names."[9] "Mutualism," "symbiotism," "symbiosis": it is the last word that stuck and that would have a long future. Frank and de Bary defined the concept of "symbiosis" as the "living together," "in common," of different species (the etymological meaning in Greek for *symbiosis*), lasting over the course of their lives (the political), within a shared external or internal habitat (the ecological). Thinking of this "living together" allowed for moving beyond Schwendener's theory, because it did not imply a mode for that shared life. For de Bary, what was important was the creation of a unity of life common to two organisms.

In the 1880s, plant roots got their turn to be analyzed from the symbiotic perspective. The fungi discovered on roots in the 1850s, first conceived as parasites, became symbiotic and were named "mychorrhizea" by Albert Bernhard Frank in 1885.

Symbiosis marked another major upheaval. Henceforth, living beings were considered much more in relationship to their environments. 1866 was also a landmark year: in addition to de Bary's discovery of lichen's dual nature, Nylander demonstrated the role of lichens as bioindicators for urban pollution, and German biologist Ernst Haeckel (1834–1919) coined the word "ecology" ("oecologia") in the scientific sense of the habitat of living beings.[10] He defined it as the science "of the relationships of organisms with the surrounding world, that is, in a larger sense, the science of conditions of existence" (it is said that Thoreau might have coined the word about ten years earlier). If symbiosis was discovered at this precise moment, it's also because scientists were beginning to think about "ecology" in biology; lichen facilitated the development of this ecological thinking.

Contrary to the English use of the term, "symbiosis" in French has since integrated an important factor: the phenomena of lasting cohabitations within living beings (its etymological meaning) must involve reciprocal benefits. We can say, in English, that symbiosis is a kind of politics; in French, that politics is mutualist.

In order to be able to include the phenomenon of parasitism in this reflection on symbiosis, I have chosen to adopt the expanded, English definition here. Hence, there can be *symbiosis without mutualism* (of the parasitic or companionate type, like pigeons or cockroaches profiting from human food scraps) and *mutualism without symbiosis* (without lasting cohabitation, like the pollination of flowers by insects or the organisms that ensure the dispersion of seeds).

1866, 2016: one hundred and fifty years later, a third partner was discovered within this symbiotic ecosystem: a basidiomycete fungus (a yeast). The lichenic couple actually hides a ménage à trois.

Since then, new microscopic components have been located in the thallus (cyanobacteria, micro-algae, micro-fungi, bacteria, amoebas, viruses) that all also contribute "actively to maintain the shared living that characterizes lichen."[11] In 2020, David Leslie Hawksworth and Martin Grube proposed this new, expanded definition of lichen: "an autonomous ecosystem formed by the interaction of an 'inclusive' fungus, an extracellular organization of one or many photosynthesizing partners, and an indeterminate number of other microscopic organisms."[12]

*

I decided to research ancient texts. After long weeks of investigation, in the library and on line, I found that in ancient Greek, contrary to the generally accepted idea, the word *sym-biosis* already existed, and well before the historian Polybius (205–123 BCE), very often mentioned as the inventor of the word.[13] It is present in Antisthenes, Aristotle, Hecataeus, and so on. Used infrequently, it already designated a shared life, literally, the act of living together, involving two spouses, two companions, or two friends. In short, it described an experience of harmonious cohabitation (the "symbiosis" of two brothers who share the same opinion, having grown up together, in Antisthenes; that of a man and woman, made for living together, in Aristotle; that of astrologers and courtiers with regard to their king, in Hecataeus and Polybius; that of father and son, again in Polybius). Moreover, the word *symbios* existed as well, in the sense of male or female "companion," "one who lives with," – a "partner," to return to Frank's word.[14]

But what is most interesting and most novel, and which once again runs counter to received ideas (which make Frank and de Bary the inventors of the word in biology, disregarding the fact that phenomena of mutual partnerships had long been observed), is that we also find in ancient Greek some instances (two) which, beginning from the primary meaning of social cohabitation, relate to the biological realm through metaphor.[15] In his *Treatise on the Intelligence of Animals,* Plutarch (46–125) mentions many cases of animals living in "society."[16] One example he traces back to Chrysippus: *pinnotheres*, a small crab that lives in shellfish.

> As for the rest that are seen to swim in shoals and to observe a mutual society, their number is not to be expressed. And therefore let us proceed to those that observe a kind of *private and particular society one with another.*[17] Among which is the pinoteras of Chrysippus, upon which he has expended so much ink, that he gives it the precedency in all his books, both physical and ethical.[18] For Chrysippus never knew the spongotera, for he would not have passed it over out of negligence. The pinoteras is so called, from watching the fish called pina or the nacre, and in shape resembles a crab; and cohabiting with the nacre, he sits like a porter at his shellside, which he lets continually stand wide open until he spies some small fishes gotten within it, such as they are wont to take for their food. Then entering the shell, he nips the flesh of the nacre, to give him notice to shut his shell; which being done, they feed together within the fortification upon the common prey.[19]

This account presents two animals described as two friends in the process of playing a trick on small fishes. The rewards are mutual, according to Plutarch: the crab gives the warning, the pina does the trapping, the two of them eat.

Such phenomena were thus observed even in antiquity, and it is striking that the word used, even if it was not yet conceptualized, is the same, one thousand, seven hundred and fifty years earlier. This *princeps* example was then passed down through the ages and became a classic in zoological and philosophical works, as well as for Philo of Alexandria (*De Animalibus* [The Soul of Animals]), Montaigne (*Essays*)[20] and Ambroise Paré, who wrote in

1582, "the pina and the pinoteras render mutual services to each other, they cannot live without one another."[21] Other scientists are inclined toward the unique benefit of the crab, like Van Beneden, who takes up the example again in 1875 to present the concept of commensalism (a partnership of organisms of different species, profitable for one of them, without endangering the other). Similarly, philosopher and naturalist Theophrastus (371–288 BCE), one of the first to name "lichens," in mentioning the various reproduction methods of the olive tree (which may seem very dated to us) expressed the relationship that the tree could have with ivy in this way:

> However all plants start in one or other of these ways, and most of them in more than one. Thus the olive is grown in all the ways mentioned, except from a twig. [...] Not that what some say that cases have been known in which, when a stake of olive-wood was planted to support ivy, it *actually lived along with it* and became a tree; but such an instance is a rare exception.[22]

The Greek word "symbiosis" was then reused, for the first time in a political sense, by German philosopher and Protestant theologian Johannes Althusius (1557–1638). In 1603, he imagined political life as structured by associations of small communities of citizens called "symbiotes." This line of thinking fell within the context of an emerging form of democracy in seventeenth-century Germany, which would serve as the source of the model for European democracy. In this sense, "symbiosis" (sometimes described as "sympathy") is only another name for the body politic ("shared life" etymologically speaking, but on the scale of society as a whole). It allows for conceiving of a form of social harmony through the reciprocity of benefits that it may involve (La Boéthie revealed to us the intentionally dependent relationship concealed behind the monarchy [...]). These links between organic bodies and political bodies were not new: Plato had already proposed them, but in a vertical and ontological dimension. That is because any concept of life, any biology, rests upon a philosophical, and thus political base.

*

For that matter, wasn't the thinking on lichenic symbiosis made possible – putting technical innovations aside (the microscope)[23] – by the fertile ideological ground of the nineteenth century?

That is because, especially in that period, the social (political) world and organic (biological) world were conceived as a single continuum, whether according to Platonic, Romantic, Hegelian, or even Darwinian views (human and other living beings are considered on the same plane). As evidence, witness the great number of words used indiscriminately for both these worlds: "societies" was used for animals and plants; "commensalism" was invented in 1874 (just before "symbiosis") to describe biological interactions; conversely, social "parasitism" was criticized in the eighteenth century, and so on. The world was seen as one, *bios*

Figure 13 © Pascale Gadon-González, *Biomorphose* (4991), 2019, gum bichromate print, with *Xanthoria* lichen, Rome, Italy, 30 x 40 cm.

like *polis* both regulated by the same laws (let us think of social Darwinism [...]).

The writings of Charles Fourier (1772–1837) constitute a particularly spectacular example. The founder of social utopian communities that he called "phalanxes" conceived the world as a great All (*Théorie de l'unité universelle* [*Theory of Universal Unity*], 1822–1823) and sought to establish ties between "natural and social things." A great lover of botany, flowers in particular (his mother was named Marie Muguet!), he demonstrated the importance of the plant world for thinking about humans, notably the human soul.[24] "A fruit, a leaf, a root, are a mirror to our souls, the play of our passions." Nature is a mystery that poets endeavor to decipher (this is the vision of Baudelaire and Caillois):

> The ancients had thus glimpsed the secret of nature, general analogy. They began with an accurate principle, but they did not know how to apply it. Their allegories were fantastic; they lacked a theory of interpretation, the art of methodically explaining the meaning of each animal, vegetable, and mineral hieroglyph.[25]

It is not surprising then to see Van Beneden presenting this analogy at the beginning of the seminal work on mutualism in biology:

> Upon close examination, one finds more than one analogy between the animal world and human society and, without looking very far, one could say that there hardly exists a social position that does not, I dare say, have its counterpart among the animals. Most of them live peaceably on the fruit of their labor and practice an occupation by which they live. But alongside these honest workers, one sees those wretches who could not get by without the aid of their neighbors.[26]

The thinking on symbiosis took place in dialogue with the philosophical and political ferment of the nineteenth century

– specifically the 1850s and 1860s – that allowed for its growth. The term "mutualism" appeared in 1828 to denote a mutual aid society of weavers in Lyon, in the context of the economic crisis that affected the Lyon silk workers beginning in 1825 ("mutualist" societies brought together workers who, in exchange for monthly dues, received aid in cases of sickness, strikes, or old age). Fourier's thinking (shaped in Lyon), socialism, and the first cooperative experiments in the 1840s, Proudhon's works on mutualism (beginning in 1845, then developed in *Du principe féderatif* [*Principle of Federation*], 1863; *Théorie de la propriété* [*Theory of Property*], 1866–1871), and Marx's critique of capitalism (the first volume of *Das Kapital* in 1867) constitute the groundwork, or at least the metaphorical (analogical) support for a new way of thinking about interactions among living beings: the dual nature of lichens was demonstrated in 1867 and mutualism was conceived in 1875.

Here, for comparison, are two definitions written by two Pierre-Josephs, the first, by Pierre-Joseph Proudhon, defining social mutualism (1871) and the second by Pierre-Joseph Van Beneden, defining biological mutualism (1875):

> A social system based on equal liberty, reciprocity, and the sovereignty of the individual over himself, his affairs, and his products; it is achieved through individual initiative, free agreement, cooperation, competition, and *voluntary association in view of defense against aggression* and the aggressor, and the protection of life, liberty and property of the non-aggressor.[27]

> Thus aid between animals is just as varied as that which is found among humans: some receive a place to live, others food, and others bed and board. We find a complete system of lodging and feeding, comparable to the best planned philozoic institutions. But if, alongside the poor, we see *others who mutually assist one another*, it would hardly be flattering to characterize

them all as parasites and commensals. We think it more just in their regard to call them *Mutualists*.[28]

This analogical thinking regarding symbiosis has been highlighted by the scientific and philosophical community: Maurice Caullery in 1922, then, more recently, Olivier Perru,[29] Mark-André Selosse,[30] and Brice Poreau.[31] In addition, I would say that new political theories and new ways of thinking (symbiotic, mutualist) about living beings have issued from the same socio-historic humus. They have come to counterbalance the view of nature as a place of competition and conflict and will continue to be intertwined and mutually supportive.

As early as the 1880s, this social dimension tended to create a gulf between symbiotic (mutualist) and evolutionary (cooperation versus competition) theorists, translating into the differences between their social (socialism and mutualism versus social Darwinism) and philosophical-political (communism versus capitalism) counterparts. Russian researchers, for example, developed a more cooperative vision of evolution.[32] Reintegrated into an evolutionist perspective, symbiosis can be caught between two opposing views: the theory of "symbiogenesis," according to which the fusion of two organisms into one is the driving force of evolution; and Darwinism, according to which organisms survive only through their descendants. This explains why the concept of symbiosis was neglected for so long in the twentieth century. It was not until the late 1970s that it was truly conceived as being in concert with evolution.[33]

Nevertheless, we must note the important pioneering efforts of Karl Fedorovitch Kessler (1815–1881), professor of ichthyology at the University of Petersburg, who, as early as 1879 (two years before his death), thought to nuance the Darwinian reading of evolution by showing even more radically the importance of mutual aid in animals, notably for reproduction: "Mutual support is as much a law of nature as reciprocal struggle is; but for the *progressive* evolution of the species, the first is more important than the second."[34] This thesis had fundamental importance for Russian political thinker Pierre Kropotkine (1842–1921); it was the foundation for his definition of human societies and his thinking on anarchist socialism.

> The animal species, in which individual struggle has been reduced to its narrowest limits, and the practice of mutual aid has attained the greatest development, are invariably the most numerous, most prosperous, and the most open to further progress.[35]

On the micro-biological level, the works of US researcher Lynn Margulis (1938–2011), during the 1960s, revolutionized and generalized the approach to organelles (the microscopic constituents of cells), like mitochondria and chloroplasts. She proposed that the eucaryotic cells that contain them are actually the result of symbiotic associations with different procaryotes. For support she relied notably on the work of Russian researcher Constantin Sergeïevitch Merejkovski (1855–1921) from the years 1900–1910, involving the symbiotic origin of chloroplasts beginning from diatoms and lichens. Thus, in the same way that the algae of lichens could be "ancient" autonomous algae, the chloroplasts present in algae, that allow for capturing light for photosynthesis, could be "ancient" bacteria: multi-scalar symbiosis. According to Margulis, symbiotic interactions would thus be the driving force of evolution, through horizontal transfer of genetic materials between bacteria (or viruses) and eucaryotic cells.

In a century, thanks notably to scientific advances allowing for better examining and better understanding infinitely small microbial life, symbiosis has been observed and analyzed in increasingly numerous configurations extending to the whole of the living world; it has become the general rule, no longer the exception; a mode, no longer a revolution.[36] In 1974, with English chemist James Lovelock, Margulis formulated the "Gaia hypothesis," according to which the planet itself is one gigantic organism functioning symbiotically, harmoniously self-regulating its components. "The whole world [...] like a giant lichen," wrote Thoreau.

*

The thinking on symbiosis now extends into the natural sciences, genetics, philosophy, economy, anthropology, and the arts.[37] Everyone is taking it up. That's the case with the splendid spider webs of Argentinian artist Tomás Saraceno (born in 1973), or the installations entitled *Symbiotic Vision*, in the Zurich Kunsthaus beginning in 2020, by the famous Icelandic–Danish artist Olafur Eliasson (born in 1967), which play with interactive art to try to illustrate the idea of a world made of interactions: in one of the halls, a screen located on the ceiling reacts to the heat of human beings as they enter. Cutting-edge research is also being done now on other symbiotic fungi, the mycorrhizae, which are located on the roots of plants and create a whole network of sugar exchanges with their symbiotic partners. Symbiosis is thus revolutionizing our practices. With plants and human bodies alike being places of symbiotic cooperation, we are now rediscovering the advantages of "companion planting" (of "vegetable guilds") once familiar to our rural ancestors. This practice consists of growing many plant species (among them notably the infamous "weeds") on the same plot of land at the same time. Similarly, dietetics now likes to take into account our microbial makeup.

Such research is multiplying and more generally testifies to the current turn, a veritable "plant turn" in thinking (philosophic, anthropologic, artistic), a return to the limelight of the plant and fungal world. From the theoretical and biological perspective, this trend began at the turn of the millennium with the pioneering works of Francis Hallé[38] and Patrick Blanc.[39] It has expanded over the last decade,[40] especially with the growing importance of fungi and mycorrhizae (mosses, as well, should not be left out).[41]

The concept of symbiosis has evolved; it has been expanded, re-rooted, and neutralized. In 1991, Lynn Margulis defined it henceforth as "a set of [ecological] interactions between nonhuman organisms," involving physical proximity, different species, and significant duration.[42] The whole ensemble creates a new, complex unit that French insect biologist Paul Nardon named a "symbiocosm" in 1995.[43] It involves the "pooling" of two or many genomes toward the goal of ensuring survival and adaptation in the environment of the newly formed entity. That is to say that here we find again the original (etymological) and more general definition of Anton de Bary's "cohabitation." English lichenologist David Cecil Smith prefers the expression "mutual interdependence" which, as Olivier Perru has shown, "empties the collective imagination of idealizations of mutual aid and synergy."[44] Lynn Margulis has criticized the mutualist vision for its anthropomorphic projections, as if symbiosis involved a sort of social contract and cost-benefit analysis between organisms. Moreover, she shows that symbiogenesis does not happen without struggle or imbalance, sometimes resulting in the death or rejection of one of the symbionts.

In recent years, the notion of *mutualist* symbiosis has been deconstructed – or stripped of illusions – by the scientific community. In 2001, David C. Smith declared that the symbiotic exchange on a nutritive level is often unilateral: for one of the partners, the cost of symbiosis outweighs the benefits.[45] The fungus synthesizes an enzyme ("permease") that acts on the membranes of the algae to facilitate the diffusion of the sugars that they contain. Similarly, in the laboratory, if a fungus is provided with dissolved sugars, it tends to suffocate the algae, as if they were no longer necessary to it. Mycorrhization can also fluctuate between mutualism and parasitism over the course of time. In November 2013, a team of German and Mexican researchers published the results of a study on the relationship between the *Pseudomyrmex ferrugineus* ant and bullhorn acacia tree of Central America. The tree provides the ant with the only type of sugar that it can assimilate in exchange for its protection against plants and herbivores. According to their conclusions, the ant larvae secrete an enzyme ("invertase") that allows them to assimilate any type of sugar, but upon contact with the nectar of the acacia, they no longer produce it, as if the tree had rendered them dependent on its own sugar. Notably because of these enzymes, their symbiosis seems to be increasingly reduced to a form of parasitism, a "master/slave relationship," Schwendener would say. In *lichens*, there are *links*, but they come with all the ambiguity of "links," which unite but also enchain. Any partnership may be imbalanced. Symbiosis, which assumes an innate horizontality among living beings, is a space of convenient and fertile projections, which also tends to be relativized – at the same time as democratic models show increasingly unilateral or "parasitic" tendencies.

> The difference between you and Talita [...] is something that is obvious to the touch. I don't understand why you have to pick up her vocabulary. I'm repelled by hermit crabs, symbiosis in all its forms, lichens, and all other parasites.
> Julio Cortázar[46]

*

It may be a matter, then, of moving beyond the concept of symbiosis by returning to its original sense, void of mutualist meanings: a suspended relationship, without a defined status, beyond cooperation and competition. As David George Haskell has written in the magnificent pages that he devotes to lichen in *The Forest Unseen: A Year's Watch in Nature*, "We need a new metaphor for the forest, one that helps us visualize plants both sharing and competing."[47] And for that, we must no longer rely on the idea of the individual:

> The lichen partners have ceased to be individuals, surrendering that possibility of drawing a line between oppressor and oppressed. Like a farmer tending her apple trees, and her field of corn, a lichen is a melding of lives. Once individuality dissolves, the score card of victors and victims makes little sense. Is corn

oppressed? Does the farmer's dependence on corn make her a victim? These questions are premised on a separation that does not exist. [...] Lichens add physical intimacy to this interdependence, fusing their bodies and intertwining the membranes of their cells [...] bound by evolution's hand.[48]

More than the nature of the symbiotic relationship, the most important thing may be this revision of the concept of the individual that symbiosis necessitates: the idea that living beings are porous, interdependent, open to "trans-species" and "trans-kingdom" alliances. Life is interstices. Life is *trouble* (Donna Haraway). Living is thus, essentially, a *politics*: the world as interspecies and interkingdom politics, like a "worldwide web" with moving, dynamic configurations, these networks participating in the evolution of the species.

Chimeras, Vampires, and Other Common Monsters

With its aberrant appearance and monstrous beauty, lichen is also a chimera.

In Greek mythology, the chimera is a monster composed of a patchwork body of different animal species. But chimeras are no longer chimerical: the lichen body is, among other things, part algae and part fungus or fungi, part plant and part fungus, part earth and part sea, at the crossroads of kingdoms and ecosystems; it is a conjunction of intensities.

Turning from a morphological to a genetic perspective, we can now speak of "chimeras" to describe organisms composed of distinct genomes: this is the special case of reproduction between different animal species or that of the plant graft, which brings together two plant species, each retaining its genome, within a single organism – and the ordinary case of lichen.

Lichen is like a "double star." It is generally perceived as a single body, but as soon as we look more closely, we see that the star we think we have distinguished in the night sky in reality hides two

Chimeras, Vampires, and Other Common Monsters

Figure 14 © Pascale Gadon-González, *Biomorphose* (5110), 2019, with *Anaptychia* lichen, originally in color, Rome, Italy, pigment print, 30 x 40 cm.

stars (principals) in mutual orbit, each turning around the other (or more precisely, around an absent center – although one of the stars may be more powerful than the other).

*

Our own bodies cohabit with thousands of bacteria. It is estimated that more than twenty thousand different species live in us, and with us, and that ninety percent of the total number of cells present in our bodies are bacterial or fungal. "Our" body is an outdated concept. They live especially in surfaces exposed to the environment (skin, nose, small intestines, and colon).

The mechanism of symbiosis invites us to redefine the boundaries of biological individuality, as well as the limits of anthropology. Substances no longer exist.[49] The individual is neither unitary nor closed, but compound, divided, and in relationships with a symbiotic retinue at once mutualist (on the order of cooperation) and parasitic (on the order of competition). The evolution of the human species is the fruit of "lichenizations."

Although the discoveries of symbiosis date back centuries and, with regard to the human body, go back to the 1960s and 1970s, it is especially over the course of this last decade that philosophy has seized upon it. After the Anthropocene, the *Lichenocene*. We can observe a growing trend for applying the lichen metaphor to human beings: launched in March 2012 by US biologist David George Haskell ("we are lichens on a grand scale")[50] and in December 2012 by US biologist Scott Frederick Gilbert ("we are all lichens"),[51] it was extended, notably by French philosophers Olga Potot (in an inclusive version: "*Nous sommes tou-te-s du lichen*" [We are *all* (feminine and masculine forms) lichen] in 2014)[52] and Karine Prévot ("*Somme-nous des lichens? Une perspective végétale sur l'individu*" [Are we lichens? A plant perspective on the individual] in 2018),[53] and yet again by the famous US anthropologist Donna Haraway in 2016 ("We are all lichens, all corals").[54] Beginning with Donna Haraway's book, US artist Laura C. Carlson reflects radically on ethical and ecological, as well as feminist and decolonialist implications of this declaration with her 2018–2019 exhibition of a series of embroidered banners entitled *We Are All Lichen*.[55] Different species of lichens are represented, enlarged, "altered," and suspended vertically, appearing as so many flags advocating and imagining a way of thinking about relationship.

> I created ten banners featuring lichen "maps." In lichen maps, I can create points of clarity within the lichen, where species mingle, where they grow, and how they endeavor together.[56]

Like coral, lichen has the peculiarity of presenting individualized components within a single structure that is reproduced as such. It is an organism *formed* by symbiosis: this is not an acquired function but a *condition* of growth. According to Karine Prévot, "that is how the question of individuality [...], and the question of symbiosis that it brings to light, clear the way toward a conception of individuality open to communities and ecosystems as a whole."[57]

The vertical and determinist vision of genetics, inherited from the twentieth century, is thus outdated. Recent research is headed in this direction: the symbiotic environment plays a role in the genetic makeup of an individual. Genes are not simply innate, some can be acquired through transfers of symbiotic partners. The Darwinian theory is thus being reevaluated; evolution is no longer linked to the individual struggle for survival, but to relational configurations that can be its driving force, as we have seen. Survival is no longer that of the fittest, but the result of the most effective interactions.

In this context, existence precedes essence; we define ourselves by our choices, but also involuntarily by our relationships with our companion species, our "symbionts." Essence is thus a shifting, relational, and ecological notion. The individual is only "the visible foam of a microbial world," Marc-André Selosse insists poetically.[58] The unicity of the pronoun "I" that designates "me," the human speaking self, which was demolished on the philosophical, psychological, and linguistic planes beginning in the nineteenth century, was something poetry had long questioned. Let us recall Rimbaud's now famous line from 1871: "I is another." Modern poets sought to alter the "I" and open it to the "you," to create a polyphony and a theatricality deconstructing the "I" into a precarious role. "I was born full of holes," wrote Michaux in 1929. This time, its unicity is being contested on the biological plane. The "I" is fundamentally open to its symbionts, a sort of fourth human "narcissistic wound," to adopt Freud's expression (the earth is no longer the center of the universe, human beings are no longer the exception among living beings, nor the "sovereigns" of their souls, nor, henceforth, of their biology).

*

Until now, not many artists have played with the metaphor of lichen's symbiosis to formulate the modern idea of a language that no longer comes only from a single subject but is an organization, an ecosystem. Montaigne characterized his essays as "chimeras." In his *Troisième Dessous* (1977), Michel Butor considers the practice of intertextuality, of textual dialogue, as a "symbiosis."

This work is made up of dreams that are visual homages to painter friends. Lichen is thus ideal for the title of the one dedicated to Saby ("The Dream of Lichens"). But more generally, Butor defines the writing in these homages, which are art critiques based on encounters between Butor's language and that of the artists in question, as a "compenetration" of styles. This intertextual practice, which involves integrating the voice of the other with one's own voice, sometimes violently, is described as a sort of "vampirism." But this act is reciprocal (the two speakers are vampires): the vampire transforms the dead even while transforming himself, in a mutual revival. This image makes me think of the "anthropophagia" movement in Brazilian modernist literature, which consisted of devouring texts to better appropriate them, to be transformed by transforming them, and *vice versa*. Butor evokes the concept of symbiosis in this way:

> "The Dream of Lichens" is a text in which I integrated an interview with a painter of my friends. And, by transforming a certain number of terms in the text of the interview, I obtained a species of vampire conversation, with all sorts of branching, to try to create a text that was itself a sort of lichen, lichen being, for me, in this whole thing, the very image of symbiosis, that is, collaboration. All these texts, made beginning from works of painters, are lichens. *Organisms that compenetrate,* to arrive at these *configurations* of lichens.[59]

This polyphony materializes in the play between italics and regular typeface, between prose and poetry. This fragmentation of the writing and how it occupies the page can call up the visual image of lichens on their supports. Butor evokes this idea a year later:

> with my entomological tweezers to thus isolate
> on the white pages gradually darkened by my glosses

a few of your iodized ink spots (September
the dahlias in the gardens classes about to begin again
and all the buzzing difficulties of this year
of the unexpected
Marie-Jo returns the car the trunk already empty)
like yellow and gray lichens on the granite
[...]
> *pages*
> *lines thighs and time*
> *spots lichens*[60]

The organic model for describing the way text functions is not new, but the symbiotic model is. It allows a dialogic ideal, both intertextual and intermedium, to be formulated, permitting reciprocal "transits" between two artists united in the ecosystem of the page.

A "Third Place"

The nineteenth-century concept of "symbiosis" (the "algo-lichen hypothesis") made it possible to describe the double nature of lichen, observed on the microscopic level. It allowed for the transition from a unified and universal concept of the biological individual (the "lichen being") to a plural and shifting definition, an ensemble of parts.

This deconstruction of lichen has resulted in its fragmentation into something like building blocks. There are the fungus, algae, yeast, and microbes that cohabit in the same structure, each with its own specialist in the scientific world. This interpretation, though necessary for better understanding lichen, has had the gradual effect of favoring the perspective of the fungus. The lichen is supposedly only a "nutritional strategy" for the fungus, a "fungal intention" in the same way as other fungi might cohabit with tree roots or decomposing matter. In 1961, this *reductionist* positioning, made possible by the microscope, resulted in the classifying of lichen within the kingdom of fungi: these were "lichenized fungi" and lichens were named according to the name of the fungal partner. Henceforth, lichen as the whole ensemble went nameless; it had no name as such but bore the name of the main fungus that was part of it. In designating the whole by one of its parts, the scientific names of lichens are, in fact, synecdochical – they have thus become partial in both senses.

This has its virtues on the level of taxonomy (the classification of species) and general understanding. Now "lichenized fungi" are integrated into one clear realm, separate from plants, and are distinguished into twenty thousand species of fungi, each with its own phylogenetic path.

Such a postulate raises certain questions however. First of all, this reductionist bias is ... reductive. Lichens are just as much algae (which are themselves plants), they bridge the wide gap between the realms. Taxonomy does not allow lichen to be classified as one entity, but only from the perspective of the fungus on a microscopic level. Moreover, the prevalence of the fungus within the symbiotic whole may be debatable. Although it represents the largest part of the biomass and often gives the lichen its form, although it may be the one who pulls the strings in this suspect mutualism (through enzymes), although, finally, it allows for more precise classification (more than twenty thousand species of fungi as opposed to one hundred algae identified within lichens), these criteria are relative. Lichen could also very easily be viewed from the perspective of the algae. Algae play a major role in how lichens function (they are responsible for photosynthesis). Furthermore, the species of lichenized algae have a greater autonomy than the fungi. These single-cell green algae can live in the absence of their mycobiont. That is the case with those of the *Trentepholia* type that are seen on the leaves of bushes (coffee trees, pepper plants) or as reddish crusts on moist walls of houses. These same algae can also participate in symbiotic associations with other living beings (liverwort, water fern). Lichenized fungi themselves cannot live alone. In a similar fashion, the photo-symbiotic partnership is necessary only for the Roscoff worm, which cannot live without its algae, whereas the algae can live alone in a marine environment.

From this perspective, the lichenic whole is no longer a "species" in the biological sense of the word, except as reduced to its mycobiont. The taxonomic notion of species does not allow for conceiving of this type of symbiotic organizations, of unities on this scale.

Nor does the reductionist perspective allow for thinking effectively about lichen holistically, for considering it as a coherent system that differs from the sum of its parts.

In his thesis published in 1964, Gilbert Simondon (1924–1989) reflected philosophically on lichenic symbiosis in his revision of the concept of "individuation" (he drew on the works of Schwendener and nineteenth-century thinking on "association," contrasting it to parasitism). He sought to reinscribe the individual in time and space: no longer as substance, monad, but as process, act, product of an environment. Fungus and algae appear reciprocally as the "external environment" of the other symbiont in the lichenic ensemble. The vital functions of each are not dissolved within the whole, because the fungus and algae are able to revive them if they are accidentally separated from the consortium. The study of symbiosis, which makes the case for organic "societies" or "colonies" (lichens, corals, sponges), allowed Simondon to demonstrate that individuality is multiple and scalar:

> Here, association constitutes a kind of second individuality superimposed over the individuality of the associated beings, without destroying it. Here, there is a reproductive system of the society as a society, a reproductive system of the Fungus as Fungus. The association does not destroy the individuality of the individuals who compose it.[61]

It is a question of conceiving lichen as well on the level of this "second individuality," of this lichenic ensemble. Lichen really is much more than a fungus that has learned how to farm; it allows both fungus and algae (and other symbionts) to develop new potential. Symbiosis is a relationship of the "additive" type. As Simondon writes, "the total quality of the organization of beings thus constituted goes far beyond that of a single individual [...], the activity of each being is conveyed through the much greater capacity for activity of the partnership, [...] which leads to an increase of capacities for the whole ensemble."[62] In effect, symbiosis cannot be reduced to the juxtaposition of the properties of the partners (1 + 1 + ...), instead, the interaction allows for a "second individuality" (1 + 1 = 1) to earn new so-called "emergent" properties (1 + 1 > 2). That is what permits lichen to develop its incredible resistance, as well as what allows numerous plant and animal species to better adapt to different diets or to protect themselves from predation through associations of microbial and fungal symbionts. The morphology of the fungus is directly altered through its association with the algae, which carries out photosynthesis. Generally composed of two layers of cortex protectors surrounding the layers of photosynthetic algae and an aeriferous zone (the medulla, where the gases circulate), the form and structure is that of a leaf. Lichen thus invented the leaf well before "higher" plants! Moreover, it is possible to cultivate in the laboratory the mycobiont by itself, but from then on, it presents a very different amorphous structure. Finally, symbiosis allows lichen to synthesize "lichenic substances" that protect the association from external aggressions. In the case of *Xanthoria parietina* (or, should I say, the lichen that includes this fungus), the parietin pigment, which gives the fungus its name and the thallus its color, offers protection from the sun and herbivores (it is toxic to them). And these substances (some seven hundred of them) are found to be practically absent from fungi that are not lichenized (only fifty of them).

Recent studies have shown that the fungi of lichens can alter their nutritional "farming" strategies over the course of their lives by changing algal partners in order to better withstand very diverse climatic conditions.[63]

A "Third Place"

As Canadian biologist Trevor Goward (born in 1952) points out, it is a matter of not remaining bound to the reductionist (Schwendernerian) approach that emerged in the nineteenth century and to thinking of lichen dialectically, on these two levels at the same time, or even beyond them.[64] The holistic level is empirical: in our experience of lichen, we see it as one. Its popular names are now the only ones to name it in its totality: parmelia designates a lichen; *Xanthoria parietina* its principal mycobiont.[65] So that is the positive side of blindness, of returning to an age before microscopes. It allows us to consider lichen from the perspective of its macroscopic morphology, and notably, to see how it can respond, by more than genetic factors (the internalist perspective), to environmental disturbances. As Trevor Goward writes, within a single "species" of lichens, there exist no two identical thalli, as each bears the trace of interactions with the environment in which it lives. It would be a matter, then, of drawing concurrently on the contributions of traditional (morphological) and modern (genetic) botany, "of authorizing two systems of nomenclature – one that targets the lichen fungus and emphasizes phylogeny more than morphology; and the other that applies to the lichen as a whole and emphasizes morphology ["phenetic" taxonomy] more than phylogeny." It is a matter of looking at reality with a double focus, the two planes nurturing one another reciprocally (the discovery of microscopic symbiosis allows for better understanding the physiology of lichen).

*

Being neither species nor individual, how then to represent lichen? Don't all species have their own retinues of symbionts? How to designate unities of living beings?

It might be tempting to apply the concept of "ecosystem" (or "micro-ecosystem") to the lichenic ensemble. This word was thus defined by English botanist Arthur George Tansley (1871–1955), pioneer in the study of plant ecology, in 1935: "the totality of the system [...] including not only the complex of organisms but also the whole complex of physical factors ... the factors of habitat in the larger sense. [...]"[66] The concept of ecosystem lets us consider lichen as a habitat, an "environment," open to its various components, as a system articulating the whole with its parts. Nevertheless, as with the concept of "superorganism," it maintains the idea of a unity of the whole, it does not allow us to describe symbiotic relationships that are located *outside* the lichenic structure itself (or outside of other living beings).

Symbiosis is an active coexistence of individualities. The morphological and genetic definition of the individual must thus be reevaluated in light of this retinue of symbionts that interact over the long term or at regular intervals with the host, in a mutualist or parasitic way. In 2002, US biologist Forest Rohwer (born in 1969) expanded upon a concept proposed by Lynn Margulis.[67] He suggested calling the ensemble composed of the host and all its micro-organisms a "holobiont" (from the Greek *holos*, "entire").[68] He began by considering coral, which is a symbiosis between an animal (the polyp) and algae (the zooxanthellae) and which brings with it as well a whole complex microbial retinue. In this sense, lichen would be the holobiont of the fungus, algae, and other tiny partners, known or not yet known, in this meta-organism. Here we are approaching what Paul Nardon called a "symbiocosm." For Marc-André Selosse, however, this model maintains the concept of organism (or unity), even if it expands it, and does not allow for taking into account the networks woven by the microbes. The same mycorrhizal fungi colonize the roots of different plants, creating an immense subterranean network of sugar exchanges in the forests. Likewise, insect pollinators do not gather pollen from a single flower, and thus design an aerial network. The holobiont concept hides the importance of the relationships between the symbionts.

Thus lichen eludes the boundary between ecosystem and organism; it constitutes a *tension*, between an ecosystem that is supposedly "crystalized," and an organism that is supposedly open. The most useful concept, it seems to me, for moving beyond this unitary vision of the living being (organism) and for taking into account the whole ensemble of these symbiotic relationships is the one that Marc-André Selosse more radically proposes: that of "interaction."

> Modern science has transposed a Western philosophy based on the individual into a biology based on the organism. A true rupture would give interactions the central place. A spider web is not an ensemble of points, but above all the threads that hold them together.[69]

This dynamic and "tensive" vision considers all living beings to be connected to a network and actually goes back to the philosophical concept of "haecceity" as defined by Gilles Deleuze and Félix Guattari in 1980. Their concept involves thinking of individuation beginning from another biological metaphor: the rhizome.

> A haecceity has neither beginning nor end, nor origin nor destination; it is always a middle. It is not made up of points, but only of lines. It is a rhizome.[...] Any individuation is not done by way of a subject or even a thing [...]. The rhizome will not be reduced to the One or the multiple. [...] It is not made up of units but of dimensions, or rather of shifting directions. It has no beginning or end, but always a middle, by which it grows and overflows. It constitutes multiplicities.[70]

An underground, horizontal stalk of certain perennial plants, which can serve as root, stalk, or branch according to its position on the plant, the rhizome allows for conceiving of a dynamic ontology that opposes any idea of hierarchy or closure. Living beings appear as configurations of intensities, as environments of interactions.

On a political level, philosopher Dénètem Touam Bona developed the metaphor of the liana in order to show the power of "allied communities," another sign of the current success of plant metaphors. According to him, "the plant model of the liana" allows us to "think about the emergence of an "Us" – a decompartmentalized community – encompassing the most diverse individuals and groups":

> The liana possesses formidable interlacing powers. Its ascent toward the sky is possible only because it relies on others. [...] The Breton word *lienaj*, the likely root of "liana" and of *lyannaj*, goes back to the making of cloth, to the knowledge of textiles.[...] The "living pillars" evoked by Baudelaire, should be conceived today more like an inextricable tangle of supple, adjustable lines that gather together, in one and the same texture, multitudes of living beings and elements, including those we call "humans." [...] The aerial tangle of lianas, just like the underground network of roots and mycelia, help to make the forest a shifting web [...], a *toutmonde* (as Glissant says) constantly being reinvented.[71]

This image, with its political and sociocultural implications, can serve as a model for thinking interactively about living beings, where we are "all interlaced" and where we are "never alone" (to echo the titles of recent scientific works), already present in the German word for lichen, *flechten* ("to weave, to braid, to interlace").

Thus a new biology of interaction has been emerging in recent years, which does not replace but nuances or completes the biology of the organism: the "ecology" is replacing the "sociology" of associations and can influence the physiology of individuals.

In this sense, lichen is not just an original megastructure (composed of two or three independent symbionts and maintained in a single additive structure); rather, it is an environment interacting on many levels. The lichen-environment reproduces as a whole (through dissemination of lichenic fragments) or in part (fungi spores), and thus recreates a different environment.

*

This "ecology" of living beings allows us to think, for example, about one of the reproductive strategies of our famous, intrepid *Xanthoria parietina*. Like most lichens, they can reproduce by dispersing through the air either fragments of the lichenic complex (fungus–algae) that are "ready-to-use" (like cuttings) or single fungus spores, momentarily interrupting the symbiotic process, that will combine through chance encounters and then reenact the symbiosis history. Spores, released by the apothecia, germinate freely while waiting to find their algae (in what is called an "aposymbiotic" phase). In this case, one strategy, more certain but less ethical, consists of the fungus spore stealing the (algae) partner of another lichen. This form of reproductive parasitism explains why *Xanthoria parietina* very often lives in the company of another species, of the *Physcia* genre, which is gray in color and contains the same algae species. Symbiosis plays out here on many levels: within the lichen and outside of it. Likewise, one thallus combines several genotypes, those of its various symbionts, but we can find many *different* genotypes for the *same* species of fungus (or algae). This might be explained through a partial fusion of the lichen with neighboring thalli.

*

Is it possible to imagine a rhizomic taxonomy that could take into account this biology of interaction? Is lichen a plant or a fungus? Roscoff worms that ingest algae and can then feed themselves through photosynthesis – are they animals or plants? Are we seeing the extinction of "species" and the end of the realm of "realms"?[72] Could anthropology be expanded to include the symbiotic retinue of humans?[73]

The boundaries between nature and culture have become blurred. The world is a place of cohabitations (symbioses) with interactions that can be seen as "cultures" (lichenized fungus cultivates algae). As Emanuele Coccia writes, these interactions are less natural than "technical," "artificial," "of a cognitive and speculative order":

The world then is this relation of reciprocal culture (never defined purely by the logic of utility, nor that of free usage). In this sense, no ecology is possible, because every ecosystem is the result of an agricultural practice and the involvement of other species. There is no wild space, just as there are no wild animals, because everything is cultivated. The relationship between culture and nature is always reversible: any species can embody nature for us, and vice versa.[74]

*

In recent years, the artistic practice of Pascale Gadon-González has explored precisely this lichenic symbiosis by adopting the mutualist idea of a balance of exchanges: "Notions of relationship, interaction, reciprocity, coexistence, and cooperation are at the origin of my artistic work."[75]

She describes this "second individuality" (on the level of symbiotic lichen) through this image of a "third place," a sort of utopia:

> My encounter with the world of lichens has crystallized a realization for me: what I observed and what I considered to be a single living being was in fact only the organization of two others that at the same time constituted it as a whole.[...] Existence, for lichen, is no longer located in a relationship of filiation, but in one of a conjunction that simultaneously forms an "other," a third place, a new "being" in the world.[76]

Beginning with the latest scientific observation techniques, and notably in collaboration with the Center for Applied Electronic Microscopy in Biology at the University of Toulouse III – Paul Sabatier, Gadon-González is seeking to reflect on this upheaval in the conception of the living being, as played out especially in lichen cells, in the smallest of the small. In *Paysages SP*, she conceals within her panoramic landscape photographs some images of lichens made from electronic microscopy scans (of the surface), which she calls "augmented landscapes." In the dead leaves on the ground or the clouds in the sky, the eye gradually comes to recognize these microscopic fragments of lichens. In this way, the work creates a hesitancy in the viewer's mind, blurs for a moment identification, nominalism.

In her spectacular images, *Conjonctions* and *Cellulaires* (2018; see Ill. 14), made this time with electronic microscopy transfers

Figure 15a © Pascale Gadon-González, *Paysage SP*, 2019, gum bichromate print, palladium plate, or pigment print, 100 x 160 cm.

Figure 15b © Pascale Gadon-González, *Cellulaire*, 2018, with *Xanthoria* lichen, gum bichromate print, 38 x 28 cm.

(sections), Pascale Gadon-González offers a subjective dream on the symbiosis of lichens by coming as close as possible to the mechanism, by zooming in precisely on the area where nutritive exchanges occur between the fungus and algae (the "points of clarity" mentioned by Laura C. Carlson).

These can take two forms: the fungus filaments (the hyphae) become applied to the surface of the algae cells either through expansion, penetrating the walls of the algae (appressorium), or by penetrating the cell directly, through what is called a "sucker" (haustorium) – symbiosis from *licking* to the *French kiss*.

Such images are made possible solely through technical advances. This symbiosis sometimes remains very suggestive, subjective, dreamlike. The images trace the touching and joining of cells; the desire in the movement of the membrane that swells "like a balloon, is directed toward, tends toward contact."[77] Superimposed on the microscopy, in the background, is a photograph of the landscape where the lichen under examination was found. Lichen as memory of place, in an image or herbarium (as with Sbarbaro), or more precisely, memory of its original symbiotic environment. As botanist Trevor Goward has pointed out, this superimposition of levels has precisely the virtue of looking at lichen through a double focus, or rather by merging two focuses: the cellular, fragmentary perspective, and the global, macroscopic context. The inlay of the landscape provides the image with colors; the photomontage allows only for the natural shades of the Charante countryside – blues, greens, ochers – where the artist

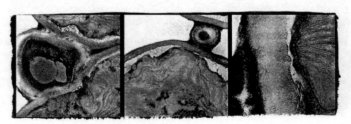

Figure 16 © Pascale Gadon-González, *Conjonctions 1*, 2018, gum bichromate print, 38 x 28 cm.

resides (La Vergne). There is no representation; the landscape is merely suggested, abstract, to make the layout and movement of the cells stand out more clearly in the foreground. This is the reverse of the process used in *Paysages SP*. Here, there is the inlay of the macroscopic into the microscopic.

In the *Bio-indicateurs* series (see Ill. 15 and 16), begun in 1998, different varieties of lichens are scanned this time, in color, on a black background and large format (often 120 x 80 cm), and identified. They appear less like a natural semiotics than symbols of life (*bios*): suspended, airborne, they stand out against the black background like lit planets emerging from the cosmos. The microcosmic and macrocosmic very often meet in Pascale Gadon-González's work: these lichens are hymns to life, to the origin of the world. The technique used, scanning with a large format camera, unlike photography, allows for multiple focal points: the view is no longer centered, hierarchical. This "symbiotic" technique lets us distinguish the lichen's details as well as its general form, the local as well as the global. The scanner is diverted from its normal use, thus allowing us to see the lichen's volume. A particular light creates a kind of chiaroscuro (moonlight? earthlight?) that comes to dramatize the apparent objectivity of the vision. The result is splendid: all the details of the complex texture and the color play of the different lichens are brought to the fore, making these photographs seem like true portraits.

In the aesthetics of symbiosis, there is also a politics at play, which Pascale Gadon-González brings to her educational and community work in Charente ("lichen as concept," as "another way of finding one's position or of being," "another relationship with the world").[78] She has created and supports several organizations and art schools in rural environments (*Le Grand Jeu*, and then *Art dans la Nature*) that seek to explore new modes of interaction.

*

The studio of Brazilian painter Luiz Zerbini (born in 1959), close to Rio's Botanical Garden, is in dialogue with the vegetation that surrounds it. In Zerbini's monumental canvases, there is this same profusion and outpouring of life, this same entanglement of nature and culture. They represent landscapes without human beings but filled with their presence (with many objects and artifacts: hydraulic pump, shoes, electric wires). Nature and culture are painted in them with great precision and in the same language (forms, colors, movements), tracing a kind of symbiotic landscape where these two concepts are no longer differentiated. This often results in a collage of travel photos or remembered images. It involves revealing the surprising colors that nature can take on, radically "pop," even "punk" and psychedelic, revealing, in short, its surreal and spiritual side.

A sign of this aesthetic of intermingling: no surface is perfectly smooth in Zerbini's paintings. There are always textures, motifs, superimpositions, impurities at work. For this, he uses rolled acrylic paint, making the paint on certain surfaces drip and run, and even resorting to "marbled paper" techniques. In *Mamanguã Reef* (2011) (see Ill. 12), bamboo stalks and seaside stones are covered with multicolored lichens – recalling especially classical Far Eastern painting (see above, pp. 86–87). Existence is not fixed, frozen, but in perpetual motion, metamorphosis, writes Emanuele Coccia. Hence the omnipresence of cables, lianas, and other branches that bind the various elements together.

Cohabitation

And yet, in the hopeful politics we seek to cultivate, we privilege heterarchy over hierarchy, the rhizomatic over the arborescent, and we celebrate the fact that such horizontal processes – lateral gene transfer, symbiosis, commensalism, and the like – can be found in the nonhuman living world. I believe that is the wrong way to ground politics. Morality, like the symbolic, emerges within – not beyond – the human. Projecting our morality, which rightfully privileges equality, on a relational landscape composed in part of nested and unidirectional associations of a logical and ontological, but not a moral, nature is a form of anthropocentric narcissism that renders us blind to some of the properties of that world beyond the human.

Eduardo Kohn, *How Forests Think*, 2013

Making lichen a model (for thinking about resistance, interaction) is fruitful and allows us to upend our conception of living beings, as it offers us clues for reflecting on our world. Nevertheless, as Eduardo Kohn notes, we can make lichen into a useful crutch, projecting our ethics or ideologies, our fantasies. It then says more about us than it says about itself. By itself, lichen is certainly of both the orders of mutualism and parasitism, horizontality and verticality, fluctuating between, or perhaps even located beyond these concepts. Lichen is not only a metaphor, not only a model, not only a word: it is also a living being. Thus, it is a question, morally, of taking into account lichen as lichen, of grounding our politics by proving ourselves to be tuned into the properties of these, radically other, living beings. As ecosystem made up of interactions, it invites us to consider this continuity between human and other living beings, this "borderless common" (Dominique Lestel). To try to learn what it means *to be lichen*. To let ourselves be transformed by it: "to become plant," exclaims US anthropologist Natasha Myers – "to become lichen."

> When I pay attention to how birds interact with water, or how mosses interact with water, or how lichens interact with water, I feel a kinship with them. I know what a cold drink of water feels like, but what would it be like to drink water over my entire body, as a lichen does?
>
> Robin Wall Kimmerer[79]

If aiming to "be" lichen as lichen, to think of lichen as lichen, to try to "make lichen speak," seems just as much an anthropomorphic trap, we can nevertheless, morally, show ourselves to be attentive to its existence, to its distinctive characteristics, to be tuned into it, and to build a moral relationship with it. As Michel Serres wrote in 1990, "rights of symbiosis are defined by reciprocity."[80] Repurposing Rousseau's expression, Serres develops the idea of a "new social contract" that expresses the necessity for a "symbiotic" relationship (in the mutualist sense) and no longer a "parasitic" one between humans and other living beings, and the environment, aimed at "minimal, collective limitations of parasitic action." Readopting Rousseau's phrase with variations shows the necessity of now inscribing any political contract into a global dimension that includes ecology, establishing nature as "legal subject," and aiming toward a "bioculture." In this sense, the symbiosis of Gaia is no longer a given, but in peril, and therefore a priority.

> An armistice contract in the objective war, a contract of symbiosis, for a symbiont recognizes the host's rights, whereas a parasite – which is what we are now – condemns to death the one it pillages and inhabits, not realizing that in the long run it is condemning itself to death too.[81]

Lichen, sentient being, one of our "cohabitants," must thus become a partner in our thinking and our life. For us, it offers a thought experiment: is it possible to locate ourselves on another scale, on its scale, that of the micro-habitats and minimal interactions that occur at this level? To conceive, for example, of its relationship to space (immobility) and to time (infinitely slow growth)? It also invites us to become aware of the neglected biodiversity and the reasons for its neglect. In cities, to adopt the mode of "flaneur," though neither that of dilettante nor modern bourgeois (Baudelaire, Benjamin), but that which offers to reconnect us with other living species who are our "cohabitants." In dialogue with lichen, we will also be more aware of the atmospheric conditions in our environments, the aerial world. From a moral and pragmatic perspective, listening to lichen also means trying to protect it. Fighting against its disappearance (as against that of another symbiotic organism, coral, in the oceans). Letting flourish the micro-habitats for which they are the pioneers, not cleaning up rocks, walls, and monuments unless the lichens are so numerous that they threaten some ancient piece of art. Not confining them to sanctuaries or museums, but letting the walls be alive, our habitat as cohabitat:

> I don't know if the lichen
> must fight with the rock?
> It doesn't shatter it
> it lives there
> it makes it habitable.
>
> Hans Magnus Enzensberger[82]

Not painting the trunks of palm trees with lime in cities. Creating a national list, in France, of species to protect. But also training the eye, creating this dialogue. In parks and botanical gardens, marking their presence and their incredible variety. Thinking like lichen allows us to know our ecosystem better, and the everyday environments of our wanderings.

ENVOI
SPORULES

"Pollen powder, lichen spores and sporules."
Saint-John Perse, *Winds*, 1945

Like many other species, the *Acarosporaceae* produce their spores inside a cylindrical structure called an *ascus*. However, each ascus produces not eight, but one hundred spores. When it rains, the pressure of the water falling on them makes the spores catapult into the air at an estimated speed of up to two hundred and fifty kilometers per hour. Many of them land close by, while others rise into the stratosphere, drifting over hundreds of kilometers.

*

Lichens are living beings located on the margins of, and in resistance to, the globalized world. They are very rarely used by large-scale industry and only sometimes on the local level. A few species are used for perfumes, a few for foods, especially by reindeer (for the moment, in any case, lichenic substances may change the situation). They are not valued (they are located outside the range of value), not wanted (a nuisance in gardens and on monuments, especially in the West), invisible, unrecognized. These shy, recalcitrant beings are equally resistant to harsh ecosystems as they are to "exploitation." It is extremely difficult to make them grow, to "cultivate" them, and in any case, their production is slow, with low yields. They lend themselves more to contemplation, to experimentation, where they prove surprisingly fertile. Considered even less than plants, linked as they are to the lowly kingdom of fungi, they are nonetheless models and mirrors for human beings.

To think about lichen is thus to think at the margins of this world and, at the same time, to look at it hard from this distance (they are our *sentries*). It is to be located at the edge to better consider and deconstruct the logic behind the "Anthropocene." Moreover, lichens grow spontaneously in peripheral spaces, forests (etymologically, where trees "are banished," "are put out"), the poles, deserts, shores and coasts, summits and fissures, as well as in the heart of our cities, developing spontaneously in the human "ruderal" spaces or creating alternative micro-habitats, those free zones won in the urban fabric, pilfered from its mineralization and its decoration.

They are at the heart of thinking about a post-Anthropocene world. Increasingly present for artists and thinkers, they sometimes escape a kind of fantasized idealism about symbiosis, rediscovered in the early twenty-first century in the context of a growing global awareness of climate change, even while constituting an ethical and poethical model of resistance and survival in a world sometimes seen through the prism of "collapsology."

Whether they depend on a mutualist or parasitic natural bond, the most important thing today may be this bond itself, the idea that, like all living beings, they are linked and compenetrate one another, and the fact that they stimulate and nurture, in the context of global warming, an ecology beyond the separate realms, as well as a way of thinking about a multi-species community and an expanded anthropology – that they invite us to rethink our "being in the world." As Donna Haraway writes:

> Corals of the seas and lichens of the land also bring us into consciousness of the Capitalocene, in which deep-sea mining and drilling in oceans and fracking and pipeline construction across delicate lichen-covered northern landscapes are

fundamental to accelerating nationalist, transnationalist, and corporate unworlding. But the coral and lichen symbionts also bring us richly into [...] non-arrogant collaboration with all those in this muddle. We are all lichens; so we can be scraped off rocks by the Furies, who still erupt to avenge crimes against the earth. Alternatively, we can join in the metabolic transformations between and among rocks and critters for living well and dying well.[1]

Lichens are a sign of life and an indicator in the toxic landscapes of our times: monocultures, genetically modified crops treated with pesticides, deforested areas, depleted soil, polluted water, invisible radioactivity and waves. Anthropologist Anna L. Tsing has clearly demonstrated that capitalism is the producer of ruins.

Their various biological characteristics, some of them discovered very recently, can be illuminating when it comes to thinking differently about our time. They are a model of subtle resistance, capable of living in varied and often disturbed ecosystems. Immobile, they are often very close to their support, even as they are fully directed toward the air. Finally, as French lichenologist Joël Boustie sums it up, "They are exactly the opposite of our society. They grow extremely slowly, a few microns per year for some, and live symbiotically."[2] Their identity is plural.

Lichen is also at the origin of the great models that let us understand our world. It allowed us to conceive scientifically of the ideas of mutualism and symbiosis (with the pivotal role, in this discovery, of the heterotrophic, lichenized, or myccorhizal fungus) before these concepts were expanded to include the whole of living beings. The years 1860 to 1870 were fundamental in the emergence of this scientific revolution that the realization of symbiotic mechanisms in living beings constituted. Since these mechanisms very often function on an infinitely small scale, the roll of microscopy was essential.

In the West, lichen has contributed to the model for a new biology no longer based (solely) on insularity, but rather on networks of interactions, "globalizing" the living, beginning in particular with the small, the microcosmic (the microbial world). It was at the origin of bio-indication. Due to its creative forms, it also contributed to earlier models, a theory of painting as well as the doctrine of signatures.

But lichen is not only a model, not only a support for projections or values. It is also a new partner in reflection; it is important to link it more closely to our own ecology. In this sense, this inquiry is also the opportunity to offer a hospitable stone (or strip of bark) to the effort to popularize lichen. An attempt to make it more well-liked, to make its familiarity more familiar. It participates in our symbiotic ecosystem, and develops unsuspected strategies of adaptation, sometimes indications of, or antidotes for, our modernity. It thus asks us to conceive of a *micro-ecology*, one of ruderal spaces, urban and rural micro-habitats, which must be protected.

Surprisingly, in France, not a single species of lichen appears on the national list of protected species; no watch list for endangered lichens exists. Lichens represent a neglected but very rich biodiversity, fundamental in the stabilization of soils, in pioneer ecosystems (they prepare these areas for subsequent plants by trapping moisture and organic debris, and then by decomposing themselves), and in the establishment of micro-habitats. They serve as ecological niche for many minuscule invertebrates at the bottom of the food chain. Thanks to their photosynthetic algae, it is estimated that when life began, about four hundred and forty-five million years ago, they produced, along with moss, nearly thirty percent of terrestrial oxygen.[3]

Thus lichen constitutes a complete world apart, an alternative and mysterious world for which the descriptions by scientists and writers, like Thoreau, Sbarbaro, Gascar, Zola, Butor, and Japanese haiku poets, remain engraved in my memory, ecopoetic approaches on lichen's level. In poetry, it mirrors existential quests and opens the way for reflection, in the form of fragmentary thoughts, "poor" aphorisms, or "salvaged" lines, like Butor's lichen of "breaths" that stubbornly survives on our sidewalks.

"Exercise of metaphysical freshness" for Bachelard, "diet" for Thoreau, lichen is a physical and spiritual practice, a cleanse or vital cure, a "life force" shining now as from the depths of time its small bathyscaphe light, its firefly beacon. It reveals, at the same time as it demonstrates, the relationships we must weave with the world that we share.

Notes

1 First Contacts

1. Victor Hugo, *L'Âne* (1857–1858), describes the origins of the world: "Dragging itself on its thousand feet, the kraken/ resembles a living rock under seaweed and lichen." Michel Butor rhymes "lichen" with "*haleines*" [breath]; Christian Dotremont rhymes it with "*rennes*" [reindeer].
2. David George Haskell, *The Forest Unseen: A Year's Watch in Nature* (New York: Penguin Books, 2013), p. 2.
3. Georges Perec, *L'Infra-ordinaire* (Paris: Seuil, 1989).
4. Qianlong, *Éloge de la ville de Moukden et de ses environs: Poème* (1743), trans. J.-M. Amiot (Paris: Tilliard, 1770), p. 36.
5. Without looking closely, it is difficult to distinguish on our sidewalks the white crustose lichen *Lecanora muralis* from chewing gum. Greek photographer Effie Paleologou (born in 1960) has tried to express the worked texture and the relationship between the mark and support of discarded chewing gum stuck to streets and sidewalks in London (*Microcosms* series, 2013). Flattened, resistant because nonbiodegradable, they appear like imprints, marking urban territory. The closely centered images play with scale: aerial view or tiny worlds? French sociologist Véronique Nahoum-Grappe (born in 1949), who has made "walks" into an ethical and political challenge, describes discarded gum this way, over the course of her Paris strolls: "There are forms that complete the décor of urban life, and that only appear while walking aimlessly. One example, the strange outlines of what remains of trampled chewing gum on the pavement. It took me a long time to see these rings, the only urban marks that neither

rain nor public authorities can clean up, quite the opposite of dog feces or graffiti. They are embedded in the asphalt and display every shade of gray: faded gray on black to dark gray on a light background. They are all different shapes. A perfect circle: is it from a heel landing hard on a wad still fresh from the mouth of a girl who aimed it at her boyfriend? On the metro platform, these gray traces of chewed gum are in perfect contrast to the "beautiful women" of the brightly lit advertising signs that prompt a desire to possess but also a dream life. Chewing gum is the smallest transgression against civility. [...] Once you begin to look at the traces of chewing gum when walking in Paris, you begin to see nothing but these indelible communion hosts [...]!" (interview with A. Diatkine, *Libération*, December 30, 2006).
6 Jean Rolin, *Traverses* (Paris: NIL, 1999).
7 In particular, the "Santo" (2006) and "La Planète revisitée" (2009–2019) expeditions in Africa that were very widely publicized.
8 Emanuele Coccia, *The Life of Plants: A Metaphysics of Mixture*, trans. Dylan J. Montanari (Cambridge: Polity, 2019).
9 Aurélien Miralles, Michel Raymond, Guillaume Lecointre, 'Empathy and compassion toward other species decrease with evolutionary divergence time', *Scientific Reports* 9:19555 (2019).
10 See Emmanuel Levinas, *Totalité et infini* (La Haye: Martinus Nijhof, 1974).
11 Sylvain Tesson, *The Consolations of the Forest: Alone in a Cabin on the Siberian Taiga*, trans. Linda Coverdale (New York: Rizzoli Ex Libris, 2013).
12 Pierre Gascar, *Le Présage* (Paris: Gallimard, 1972). Although it is valid to recognize the current lack of uses for lichen in modern France, there exists a whole secular history (studied by "ethnolichenology") of its uses (including uses for dyes, perfumes) that does not date as far back as the doctrine of signatures; Gascar acts as though he has forgotten it here.
13 Ibid.
14 William Shakespeare, *The Comedy of Errors* (1592–1594), 2.2.172–179.

15 William Shakespeare, *Titus Andronicus* (1594), 2.3.91–97.
16 Docteur Louis Brocq, *Des lichénifications de la peau et des névrodermites* (Paris: Société d'édition scientifique, 1891).
17 Joris-Karl Huysmans, *L'Oblat* (Paris: Stock, 1903) [italics mine].
18 Émile Zola, *Lourdes*, trans. Ernest A. Vizetelly (Amherst, NY: Prometheus Books, 2000), p. 130 [italics mine].
19 Lorna Crozier, *Small Mechanics* (Toronto: McClelland & Stewart, 2011), p. 9.
20 Lichens are not the only organisms that call to mind various kinds of hair. That is also the case with many mosses and ferns, like maidenhair ferns (*capillaris*, Latin for "hair") or trichomes (from *thrix*, Greek for "hair"), epiphytic plants, like the famous *barba de viejo* in Latin America (*Tillandsia usneoides*), and certain flowering plants, like the herniaria hirsuta (hairy rupturewort).
21 Joris-Karl Huysmans, *The Damned*, trans. Terry Hale (London: Penguin Books, 2001), pp. 146–147.
22 I discovered only the astonishing survey by Valentina Pavlovna Wasson and Robert Gordon Wasson, *Mushrooms, Russia and History* (New York: Pantheon Books, 2 vols, 1957) (Lévi-Strauss offered a review of it in *L'Express*, April 10, 1958). This monumental study, which, in the 1950s, invented the discipline of "myco-ethnology," examines the cultural aspects of mushrooms throughout history and cultures. It comes to the surprising conclusion – less so when seen in the context of the cold war – of an opposition between "mycophilic" Slavic peoples and "mycophobic" Anglo-Saxon peoples. In this work, there is no discussion of *lichenized* mushrooms.
23 Gascar, *Le Présage*, p. 23.
24 Toby Spribille, Veera Tuovinen, Philipp Res, Dan Vanderpool, Heimo Wolinski, M. Catherine Aime, Kevin Schneider, Edith Stabentheiner, Merje Toome-Heller, Göran Thor, Helmut Mayrhofer, Hanna Johannesson, and John P. McCutcheon, 'Basidiomycete yeasts in the cortex of ascomycete macrolichens', *Science* 353.6298 (July 29, 2016), pp. 488–492.
25 This case of mycorrhizal symbiosis is at the heart of current

research. It was the subject of very recent work on "plant communication," made famous especially by Peter Wohllenben and his *The Hidden Life of Trees: What They Feel, How They Communicate, Discoveries from a Secret World*, trans. Jane Billinghurst (Vancouver and Berkeley: Greystone Books, 2016).
26 Gascar, *Le Présage*, pp. 109–110.
27 Camillo Sbarbaro, *Licheni: Un campionario del mondo* (Florence: Vallecchi, 1967). *Les Lichens* 3, in *Copeaux, suivi de Feux follets*, trans. J.-B. Para, (Sauve, Gard: Clémence Hiver, 1991), pp. 57–58. With the exception of this first paragraph, the Italian-to-French translations cited here of Sbarbaro's text are from this translation [English translations are not available].
28 Gascar, *Le Présage*, pp. 26–27.
29 In the Innu-aimun language, the word *uapitsheushkamik* means "lichen of the reindeer" (or of the caribou, in Canada).
30 Joséphine Bacon, *Un thé dans la toundra: Nipishapui nete mushuat* (Montreal: Mémoire d'encrier, 2013), pp. 16–17. Papakassiku and Missinaku are two animal divinities. The first rules over the caribou, the second over the aquatic animals [English translation is not available].
31 Joséphine Bacon also links lichen to clothing and especially to ceremonial dress, which she dons to sing of the elements of her native ground: "my dress is called/ lichen, she writes in *Nous sommes tous des sauvages* (Montreal: Mémoire d'encrier, 2011). In German-speaking Switzerland, in Urnäsch in the Appenzell valley, villagers celebrate Saint Sylvester's Day by dressing up in natural materials. One group of these "Silvesterklausen," the *Schö-Wüeschte* ("beautiful-ugly") wear extremely elaborate costumes made from hay, straw, fir bows, holly, ivy, bark, moss, and lichen. In Austria, the Schleicherlaufen Festival, held every five years in the Tyrolean village of Telfs, celebrates the end of winter with a parade that brings together the "Wilden Männer" (Wild Men), traditional carnival figures since the fourteenth century whose impressive clothes are fully covered with bearded lichens.
32 Natasha Kanapé Fontaine, extracts from the poems "La Marche"

and "La Cueillette" in *Blueberries and Apricots*, trans. Howard Scott (Toronto: Mawezi House, 2018).

33 Natasha Kanapé Fontaine, *Tetepiskat* (2016); on line: generations150.onf.ca/tetepiskat.

34 Hiroshige II, *Iwatake Mushroom Gathering at Kumano in Kishi Province*, print from the series *One Hundred Famous Views in the Various Provinces*, 1860.

35 Gascar, *Le Présage*, pp. 101–102.

36 Recipe published on line: liafaydjam.blogspot.com/2007/04/umbilicaria-tripe-de-roche-et-autres.html.

37 Trevor Goward, 'Twelve Readings on the Lichen Thallus', *Evansia*, The American Bryological and Lichenological Society (2009–2012).

38 Mourning Dove, "How Coyote happened to make the black moss food" in *Coyote Stories* (Caldwell: Caxton Printers Ltd, 1933), pp. 119–126.

39 For a fairly complete summary of the medical uses of lichens on the various continents, see Carlos Illana-Estebán, 'Líquenes usados en medicina tradicional', *Boletín de la Sociedad Micológica de Madrid* 36 (2012), 163–174; and Paolo Modenesi, *Il sapore e il colore dei licheni: Una guida agli usi* (Genoa: Genova University Press, 2015).

40 Charles Baudelaire, *Fragments posthumes*, 92 : "Hygiène, conduite, méthode." *Intimate Papers from the Unpublished Works of Baudelaire*, trans. Joseph T. Shipley, in *Baudelaire: His Prose and Poetry*, T.R. Smith (ed.) (New York: Boni and Liveright, 1919).

41 Kouo Yu, "Longue nostalgie," in *Anthologie de la poésie chinoise classique*, trans. P. Demiéville (Paris: Gallimard, 1962) p. 428.

42 Michael Ettmüller, *Dissertationem medicam: De chirurgia infusoria* (Leipzig: Nicolaus Scipio, 1668).

43 Alexandre Acloque, *Les Lichens: Étude sur l'anatomie, la physiologie et la morphologie de l'organisme lichénique* (Paris: Baillière, 1893).

44 Pliny the Elder, *Historia naturalis. Natural History by* Pliny, trans. H. Rackham, W.H.S. Jones, D.E. Eichholz (Cambridge,

Massachusetts: Harvard University Press, 1938), vol. 6, bk. 26:10.
45 Pierre Joseph Amoreux, Georg Franz Hoffmann, and Pierre Rémi Willemet, *Mémoires sur l'utilité des lichens dans la médecine et dans les arts* (Lyon: Piestre & Delamollière, 1787).
46 Orcein lichen was used, beginning in prehistoric times and antiquity (Theophrastus said that it rivaled murex), and particularly in Europe in the fourteenth to nineteenth centuries, to obtain shades in the range of crimson, mauve, amaranth red, and purple. The species of the genus *Roccella* (*orchillas* in Spain) were collected on the ground and cliffs of Mediterranean and Atlantic islands, especially the Canary Islands (Lanzarote), or else imported from African colonies (Cape Verde, Senegal, Angola, Madagascar), or even other continents (India, Chili). Intended for continental commerce, the work was sometimes done at the risk the lives of those men and women collectors (called *orchilleros* and *orchilleras* in the Canary Islands – originally native *guanches*, then Christian colonists). It was at the heart of a major industry and trade, long monopolized by Catholic kings. The parelle lichens of Auvergne (many species were used, essentially *Lepra aspergilla*, *Ochrolechia parella*) were also collected in the Massif Central and used by dyers in Lyon and Paris.
47 The legendary long red beard of Frederick Barbarossa, Holy Roman Emperor (1122–1190), was said to have continued growing after his death. Enzensberger relates this mythic name, through analogy, to bearded lichens (usneas) and to the dyes, also sometimes red, that were made using certain lichens.
48 Hans Magnus Enzensberger, *Blindenschrift* (Frankfurt: Suhrkamp, 1965), extracts from the poem, "Lichenology."
49 Eugène Guillevic, *Du domaine* (Paris: Gallimard, 1977).
50 Elizabeth Bishop, *A Cold Spring* (Boston: Houghton Mifflin, 1955), extract from the poem, "The Shampoo."
51 C. Roullier, M. Chollet-Krugler, E.M. Pferschy-Wenzig, A. Maillard, G.N. Rechberger, B. Legouin-Gargadennec, R. Bauer and J. Boustie, "Characterization and identification of mycosporines-like

compounds in cyanolichens: Isolation of mycos-porine hydroxy-glutamicol" *Nephroma laevigatum Ach. Phytochemistry*, 72 (2011): 1348–57.

52 Pierre Chantraine, *Dictionnaire étymologique de la langue grecque* (Paris: Klincksieck, 1968), p. 629. We also find the Greek verb λειχηνίαω ("to cover oneself with lichen") used by Theophrastus to describe olive trees.

53 Gascar, *Le Présage*, p. 147.

54 Robin Wall Kimmerer, *Braiding Sweetgrass* (Minneapolis: Milkweed Editions, 2013), p. 273.

55 Sbarbaro, *Les Lichens*, 3, pp. 57–66.

56 Coccia, *The Life of Plants*, p. 18.

57 Michel Tournier, *Friday*, trans. Norman Denny (New York: Doubleday and Company, 1969).

58 Caves and mossy rocks are commonplace in Latin elegies and bucolic verse; in Catullus, Lesbia describes "Voluptuousness" stretched out "on those beds of moss and fern."

59 Alice Munro, "Lichen," in *The Progress of Love* (New York: Alfred A. Knopf, 1986), p. 39.

60 Ibid., pp. 41–42.

61 Ibid., p. 55.

62 Jean-Paul Gavard-Perret, article appearing in the journal *Traversées* 88 (2013) (on the subject of the book by Paul-Armand Gette, *Ma propension au débordement* (Saint-Benoît-du-Sault: Tarabuste, 2013)).

63 Paul-Armand Gette and Tenebria Lupa. Internet publication (http://www.paularmandgette.com) entitled "Les Chroniques de Tenebria Lupa," which then appeared in Paul-Armand Gette, *Dessins suspendus* (Dijon: Les Presses du Réel, 2021).

64 Gascar, *Le Présage*, p. 157.

2 To Describe, Name, Represent

1 Interview with the author, March 30, 2020.

2 Sbarbaro, *Les Lichens*, 6–7, pp. 57–66.

3. Louis de Jaucourt, s.v. 'Lichen', in Denis Diderot and Jean d'Alembert (eds.), *Encyclopédie* (1765).
4. Nicolas Jolyclerc, *Phytologie universelle ou Histoire naturelle et méthodique des plantes* (Leipzig: P. Wolff, 1799), vol. 3.
5. Georges Cuvier, Jean-Baptiste de Lamarck, Antoine-Laurent de Jussieu et al., *Dictionnaire des sciences naturelles* (1816–1845), vol. 36.
6. Blog of Richard Bernaer, 'Fonge et florule', March 21, 2016: fongeflorule.wordpress.com/2016/03/21/beaute-grecque/.
7. Barry Lopez, *Arctic Dreams* (New York: Vintage, 1986), pp. 228–229.
8. Sbarbaro, *Les Lichens*, 6–7, pp. 57–66.
9. Henry David Thoreau, journal entry from December 6, 1859, 'To Walden and Baker Bridge, in the shallow snow and mizzling rain', Bradford Torrey and Francis H. Allen (eds.), *The Journal of Henry D. Thoreau* (New York: Dover Publications, 1962), vol. 2, p. 1554.
10. Guy G. Nearing, *The Lichen Book* (Ridgewood, NJ: self-published, 1947). Sbarbaro writes:
"Lichen imitates all sorts of manufactured objects and products: waxes, inlaid works, mosaics; felt and fine fabrics; buckles and buckle tongues, cups, clubs, helmets, shields, nails; matches; ribbons, nets, fans; stamped leather; velvets and bobbin lace."
(*Les Lichens*, 6–7, pp. 57–66).
11. See Roger Caillois, *Dessins ou desseins*, first published in *Preuves* 100 (June 1959).
12. Jurgis Baltrušaitis, *Les Perspectives dépravées: Aberrations. Quatre essais sur la légende des formes* (Paris: O. Perrin, Jeu savant, 1957), vol. 1. Read the very beautiful pages in the chapter on the practice of *"pierres imagées,"* illustrated by the magnificent paintings of Mathieu Dubus (1590–1665) and Johann König (1586–1635).
13. André Breton, *Le Surréalisme même* 3 (Autumn 1957).
14. This technique may call to mind the "splashed ink" or "broken ink" (*pomo xianren*) paintings of Chinese art. In the eighth century, Wang Mo worked with ink by stamping on it, rubbing it

between his hands, or by sweeping it with his brush and his hair, often after getting drunk on wine; observing the configurations of the splashes on the silk, he then made mountains, streams, etc. emerge from them.

15 Related by Henri Amic in *George Sand: Mes souvenirs* (Paris: Calmann-Lévy, 1891), p. 47.
16 This famous text by Leonardo da Vinci was published posthumously in 1651. See Leonardo da Vinci, *Treatise on Painting*, trans. P. Philip McMahon (Princeton: Princeton University Press, 1956).
17 Giorgio Vasari, *Le Vite de' più eccellenti architetti, pittori et scultori italiani, da Cimabue insino a' tempi nostri* (1550). *Lives of the Most Eminent Painters Sculptors and Architects*, trans. Gaston C. DeVere (London: MacMillan and Company, 20 vols, 1912–1915), vol. 4, p. 127.
18 Tomas Tranströmer, "Baltics" in *Östersjöar* (Stockholm: Albert Bonniers Förlag, 1974). *Bright Scythe: Selected Poems* trans. Patty Crane (Louisville: Sarabande Books, 2015), p. 67.
19 Regarding the notions of "oblique" and "diagonal" sciences as defined by Roger Caillois, see "Méduse et Cie," first published in *La Nouvelle Revue Française* 76 (April 1959); and *Obliques* (Paris: Stock, 1975).
20 Pablo Neruda, "Lichen on Stone" in *Las Piedras Del Cielo* (Buenos Aires: Losada, 1970). *5 Decades: Poems 1925–1970 Pablo Neruda*, trans. Ben Belitt (New York: Grove Press, 1994), pp. 410–411
21 Enzensberger, *Blindenschrift*, extract from the poem, "Lichenology."
22 Jean-Baptiste de Lamarck, *Encyclopédie méthodique: Botanique* (Paris: Henri Agasse, 1789).
23 Michel Serres and Nayla Farouki, preface to *Paysages des sciences* (Paris: Fayard, 1999), p. lviii.
24 Honoré de Balzac, "Seraphita" *La Revue de Paris* (1834).
25 A.Titolo and Alberto Salvadori (eds.), *Claudia Losi: La coda della balena e altri progetti (1995–2008)* (Florence and Prato: GliOri, 2008), p. 25.

26 Gaston Bachelard, *The Poetics of Space*, trans. Maria Jolas (New York: Penguin Books, 2014), pp. 153–155.
27 Oscar Furbacken, *StorstadsLav* [*Urban Lichen*], 2009.
28 Oscar Furbacken, statement on line: www.oscarfurbacken.se/inandabout.html.
29 Gascar, *Le Présage*, p. 71.
30 A jib is a telescopic crane to which the camera is attached.
31 Oscar Furbacken, statement on line: www.oscarfurbacken.se/expeditionbeyond.html.
32 Bachelard, *The Poetics of Space*, pp. 155, 161.
33 Lamarck, *Encyclopédie méthodique: Botanique*.
34 Jean Giono, *Le Chant du monde* (Paris: Gallimard, 1934).
35 Yves Chaudouët, documents from La Criée – Centre d'art contemporain de la Ville de Rennes, during the artist's residency; "Battre la campagne" season directed by Sophie Kaplan, 2014–2015: www.reseau-dda.org/en/editorial-productions/residency-logbook/item/1676-chaudouet-yves-carnet-criee/1676-chaudouet-yves-carnet-criee.html.
36 Yves Chaudouët, *Lichens*, seven lithographs with a short text by Juliette Chemillier (Paris: Éditions Atelier Clot, 1998).
37 Gascar, *Le Présage*, pp. 24–25.
38 Bernard Saby, 'L'Usnea diplotypa Wain, en forêt de Fontainebleau', *Revue bryologique et lichénologique* 15 (1946), pp. 201–202 : "This species had not been noted in France when, in June 1946, we encountered a single but thriving clump of it in the forest of Fontainebleau, on the northern slopes of Grand Mont Chauvet. It was growing on plant debris, covering the horizontal part of a big block of semi-shaded sandstone."
39 Michel Butor, *De la distance* (Rennes: Ubacs, 1990), pp. 68–69.
40 Armand Gatti, *Les Analogues du réel* (Toulouse: L'Éther vague, 1988).
41 Michel Butor, *Matière de rêves*, vol. 3: *Troisième dessous* (Paris: Gallimard, 1977).
42 Michel Butor, "Rumeurs de la forêt: Pour André Villers," "Ombrages," in *Explorations* (Lausanne: Éditions de l'Aire, 1981).

43 Michel Butor, *L'Horticulteur itinérant* (Paris: Melville, 2004).
44 The first trace of this friendship, to my knowledge, is Butor's 1961 interview with Saby (continued two years later on the occasion of a Paris exhibition of the painter's work at the Galerie de L'Oeil: 'Conversation dans l'atelier avec Bernard Saby', *L'Œil* 79–80 (July-August 1961), pp. 37–42. Many conversations would follow.
45 *Voyage avec Michel Butor*, interviews with M. Santschi (Lausanne: L'Âge d'homme, 1985) [italics mine].
46 Michel Butor, *L'Emploi du temps* (Paris: Minuit, 1956); in the city of Bleston-on-Slee in England, invented by Butor, we also find a "Lichen Street," located "in that horrible fifth district."
47 Michel Butor, *La Modification* (Paris: Minuit, 1957).
48 Michel Butor, *Brassée d'avril* (Paris: La Différence, 1982).
49 Michel Butor, "Pause, matin, midi, soir," for Anne Walker, "Avec vingt peintres (suite)" in *Rémanences* 6 (April 1996), pp. 20–21.
50 This type of room is designed with walls that absorb sound or electromagnetic waves; a deafening silence reigns there.
51 John Cage, 'An Autobiographical Statement' (1989), in R. Kostelanetz (ed.), *John Cage Writer* (New York: Limelight Editions, 1993).
52 Between 1983 and 1985, John Cage composed a musical piece on Kyoto's famous Ryōan-ji rock garden.
53 Consider his *17 Drawings by Thoreau* (1978), based on the small rough sketches in Thoreau's journals.
54 John Cage, *John Cage Diary: How to Improve the World (You Will Only Make Matters Worse)*, J. Biel (ed.) (New York: Siglio Press, 2015, 2019).
55 John Cage, 'Music Lovers' Field Companion' (1954), in *Silence: Lectures and Writings by John Cage* (Hanover, New Hampshire: Wesleyan University Press, 1961), p. 274.
56 Cage, 'An Autobiographical Statement'.
57 Account published on this mysterious website, with no clearly identifiable author: quadrust.com/a-celebrating-john-cage-concert-from-sfemf/. This site, consulted in 2020, is inactive as of 2022.

58 See Marc Chemillier, 'György Ligeti et la logique des textures', *Analyse musicale* 38 (2001), pp. 75–85.
59 Iannis Xenakis, 'Esquisse d'autobiographie' (1980), in Gérard Montassier, *Le Fait culturel* (Paris: Fayard, 1980).
60 Let us think here of *Ionisation* (1929–1931) by Edgar Varèse.
61 Véronique Brindeau, *Louange des mousses* (Arles: Philippe Picquier, 2012), p. 108.
62 André Leroi-Gourhan, *Pages oubliées sur le Japon* (1939) (Grenoble: Jérôme Millon, 2004), p. 250.
63 Haskell, *The Forest Unseen*, p. 99 [italics mine].
64 Li Po, "Ch'ang-Kan Village Song" in *The Selected Poems of Li Po*, trans. David Hinton (New York: New Directions, 1996), pp. 12–13. "La ballade de Tch'Ang-Kan" (eighth century), in *Anthologie de la poésie chinoise classique*, p. 221.
65 Anonymous, "Your Reign" ["Kimigayo"], ninth century.
66 Jun'ichirō Tanizaki, *In Praise of Shadows*, trans. Thomas J. Harper and Edward G. Seidensticker (Sedgewick, Maine: Leete's Island Books, 1977), p. 10.
67 Kiang Tsong, "Complainte de gynécée" (sixth century), in *Anthologie de la poésie chinoise classique*, p. 167.
68 Tanizaki, *In Praise of Shadows*.
69 Yasunari Kawabata, *The Old Capital*, trans. J. Martin Holman (Berkeley: Counterpoint Press, 1987, 2006), p. 2.
70 Yasunari Kawabata, *Beauty and Sadness*, trans. Howard S. Hibbett (New York: Vintage International, 1996), p. 53.
71 Wang Wei (699–761) "Deer Park" in *The Selected Poems of Wang Wei*, trans. David Hinton (New York: New Directions, 2006), p. 40. [See also Eliot Weinberger, *19 Ways of Looking at Wang Wei* (New York: New Directions, 2016), which includes two dozen translations of this poem.]
72 This haiku is sometimes attributed to Yamaguchi Sodō (1642–1716), sometimes to Shūson Katō (1905–1993).
73 Hirai Shôri, 2016.
74 Philippe Jaccottet, *Israël: Cahier bleu* (Saint-Clément-de-Rivière: Fata Morgana, 2004).

75 On this subject, see especially Maggie Bickford, *Bones of Jade, Soul of Ice: The Flowering Plum in Chinese Art* (New Haven: Yale University Art Gallery, 1985).
76 See Michel Butor, *Transit A* (Paris: Gallimard, 1993).
77 Gustav Klimt (1862–1918) was very influenced by Far Eastern art, in vogue at the turn of the century. His birch forests, created in 1902–1903, also play with the textures of the tree trunks (moss and black lines stand out like decorative elements on the smooth, white bark of the birches), as well as with the contrast between the green of the moss on the trees and the red of the leaves on the ground, heightened by the perspective.
78 Wang Changling (698–756).
79 Haku Rakuten (772–846).

3 Ecopoetics: Life Force and Resistance

1 Jean-Jacques Rousseau, *The Confessions of Jean Jacques Rousseau* (1765–1767), bk. 5, trans. Anonymous (London, 1903).
2 Auguste Pyrame de Candolle, *Histoire de la botanique genevoise* (Geneva: J. Barbezat, 1830), p. 19.
3 Chrétien-Guillaume de Lamoignon de Malesherbes (1721–1794), magistrate, botanist and French statesman.
4 Jean-Jacques Rousseau, 'Cinquième Promenade' in *Reveries of a Solitary Walker*, trans. Peter France (London and New York: Penguin Books, 1979), p. 84.
5 An important popularizer, Rousseau had only a rudimentary knowledge of botany. Pierre Gascar said of him that he had "a boarding school girl's conception of botany" (*Genève* (Seyssel: Champ Vallon, 1984), p. 54).
6 Jean-Jacques Rousseau, 'Sur les liliacées', August 22, 1771, in *Letters on the Elements of Botany to a Lady*, trans. Thomas Martyn (London: J. White, 1785), p. 19.
7 This herbarium dates from 1769–1771 and appears in the form of a small bound volume, 155 x 95 mm, composed basically of 53 pages on which mosses and lichens are displayed.

8 Jean-Jacques Rousseau, *Lettres à Monsieur de M***: Sur la formation des herbiers* (1771).
9 Georges Perec, text published in *Cause commune* 5 (February 1973), and reprinted in *L'Infra-ordinaire*, p. 11. 'Approaches to What?' in *Species of Spaces and Other Pieces*, trans. John Sturrock (London and New York: Penguin Books, 1977), p. 210.
10 Jean-Jacques Rousseau, 'Septième Promenade' in *Reveries of a Solitary Walker*, trans. Peter France, pp. 117–118.
11 Geneva's Physics and Natural History Society was created in 1791, and its Botanical Gardens in 1817 by Auguste Pyrame de Candolle. It is worth noting that the first botanical works were devoted to matters of descriptive botany and taxonomy, the order and inventory of the natural world.
12 *La Nouvelle Héloïse* by Rousseau and pastoral poetry by Albrecht von Haller strongly influenced young Swiss botanists. The works of Patrick Bungener of Geneva's Botanical Conservatory address precisely these matters. See 'La place de Rousseau dans la tradition botanique genevoise', *Annales Jean-Jacques Rousseau* 52 (2014), pp. 249–269.
13 On this subject, see Cléopâtre Montandon, *Le Développement de la science à Genève aux XVIIIe et XIXe siècles* (Vevey: Delta SA, 1975); Jacques Trembley (ed.), *Les Savants genevois dans l'Europe intellectuelle du XVIe au milieu du XIXe siècles* (Geneva: Éditions Journal de Genève, 1987); René Sigrist, *L'Essor de la science moderne à Genève* (Lausanne: Presses polytechniques et universitaires romandes, 2004).
14 Nicolas Bouvier, *La Suisse est folle: Proposition déraisonnable accompagnée de quelques images* (Geneva: Héros-Limite, 2019), p. 45.
15 Ibid., p. 42.
16 Gascar, *Genève*, p. 49.
17 Ibid., p. 12.
18 Bouvier, *La Suisse est folle* [...], p. 43.
19 John Ruskin wrote in his autobiography: "I was interested in everything, from clouds to lichens" (*Praeterita*, 1885).

20 John Ruskin, *Modern Painters* (London: Smith, Elder & Company, 1860), vol. 5.
21 The Paradisia Alpine Botanic Garden in Valnotey, located in the Aosta Valley in Italy at an altitude of 1700 meters, features some sixty species of lichens. Hervé Cochine has been creating lichen gardens in France for several years.
22 In a preface to *La Mare au diable* (1973), Léon Cellier describes Sand as the "spiritual daughter of Rousseau." On this subject, see Christine Planté, *George Sand – Fils de Jean-Jacques* (Lyon: Presses universitaires de Lyon, 2012).
23 George Sand, letter to Flaubert, May 30, 1867.
24 Jules Néraud, letter to Éverard, April 11, 1835.
25 George Sand, *Consuelo*, trans. Fayette Robinson (Philadelphia: T.B. Peterson & Brothers, 1870), pp. 247–248.
26 George Sand, *Spiridion*, trans. Patricia J.F. Worth (Albany: State University of New York Press, 2015), p. 164.
27 I will not go so far as to claim, however, as does the US biologist Hans Bergmann, that "Thoreau's study of Linnaeus' "poor trash" of vegetation became for him a way to remake his already well-developed attention to nature into a more precise as well as imaginative practice. [...] Thoreau found lichens a particularly appropriate subject for his new writing: they are quotidian, ordinary, often completely overlooked" (paper presented at the Shaw Conference Center, July 27, 2015, entitled "'I Have Become Sadly Scientific': Henry David Thoreau's Lichenology").
28 Thoreau, *Journal*, February 22, 1852.
29 Ibid., February 5, 1853.
30 Ibid., December 31, 1851.
31 Ibid., March 5, 1852.
32 Ibid., November 24, 1858.
33 Ibid., February 7, 1859.
34 Ibid., November 16, 1850.
35 Ibid., January 7, 1851.
36 Erasmus Darwin, *The Loves of Plants* (1791), Canto 1, lines 349–350.

37 William Wordsworth, "The Thorn" (1800).
38 Wilhelm Nylander, "Les Lichens du Jardin du Luxembourg," *Bulletin de la Société botanique de France*,13 (1866), pp. 364–372 [italics mine].
39 Let us remember that, scientifically speaking, lichen is not a plant.
40 "Many young examples of foliose lichens belonging to seven moderately poleotolerant species [tolerant to pollution] were observed in 1990 in the Luxembourg Garden in Paris. No foliose lichen had been seen there since 1896 and their return is the result of the drop in SO_2 rates in the air between 1983 and 1988," Marie-Agnès Letrouit-Galinou, Mark Seaward and Serge Deruelle, "À propos du retour des lichens épiphytes dans le jardin du Luxembourg (Paris)," *Bulletin de la Société botanique de France, Lettres botaniques*, 139 (2014): 115–126.
41 "L'Amour par terre" appears in Verlaine's *Fêtes galantes* (1869). Émile Zola, *Abbé Morret's Trangression*, trans. E.A. Vizetelly (New York: Mondial, 2005), p. 102.
42 Arthur Rimbaud, "The Drunken Boat" in *Arthur Rimbaud, Complete Works, Selected Letters* trans. Wallace Fowlie (Chicago: University of Chicago Press, 2005).
43 Gustave Flaubert, *The Temptation of Saint Anthony*, trans. Lafcadio Hearn (New York: The Modern Library, 2001), p. 182.
44 Gillian Kidd Osborne, 'Plant Poetics & Beyond: Lichen Writing', *Entropymag* (September 17, 2019); on line: entropymag.org/plant-poetics-and-beyond-lichen-writing/.
45 Herman Melville, *Weeds and Wildings, Chiefly: With a Rose or Two* (1891) (published posth. 1924) (Pittsfield, MA: Melville Press, 2016).
46 Anna Akhmatova, "Secrets of the Trade" (1940), trans. Jo Ann Clark *The Paris Review* 141 (Winter, 1996).
47 Peter Hutchinson, *Dissolving Clouds: Writings of Peter Hutchinson* (Provincetown, RI: Provincetown Art Press, 1994).
48 Ibid. See also 'Notes pour un carnet de croquis', *Artitudes international* 5 (June-August 1973); on this artist, in French, see the

work edited by one of his scholars, Gilles A. Tiberghien, *Peter Hutchinson* (Lyon: Fage Éditions, 2016).

49 Hutchinson, 'Molds and Fungi' (1987) in *Dissolving Clouds*.

50 According to interviews conducted by the journalist Ferdinando Camon (*Il Mestiere di poeta* (Milan: Garzanti, 1982), it was a book by Rolf Santesson (*Foliicolous lichens: A revision of the taxonomy of the obligately foliicolous, lichenized fungi* (Uppsala, 1952), vol. 1), based on studies done by the Sbarbaro who did not like to keep books in his house, including those he wrote.

51 Camillo Sbarbaro, *Fuochi fatui* [Will o' the Wisp] (Milan: All'Insegna del Pesce d'Oro, 1956–1958). In *Copeaux*, trans. J.-B. Para.

52 Sbarbaro, *Licheni. Les Lichens* 1 in *Copeaux*, trans. J.-B. Para, pp. 57–66.

53 Camillo Sbarbaro, *Trucioli* [Wood Shavings], (Florence: Vallecchi, 1920). In *Copeaux*, trans. J.-B. Para.

54 Sbarbaro, *Fuochi fatui*, in *Copeaux*, trans. J.-B. Para.

55 Sbarbaro, *Licheni. Les Lichens* 2 in *Copeaux*, trans. J.-B. Para, pp. 57–66.

56 Interviews with Camillo Sbarbaro by the journalist Ferdinando Camon, *Il Mestiere di poeta*.

57 Sbarbaro, *Licheni. Les Lichens* 2 in *Copeaux*, trans. J.-B. Para, pp. 57–66.

58 In Sbarbaro, *Fuochi fatui*. In *Feux follets*, trans. P. Gabellone, in Camillo Sbarbaro, *Pianissimo, suivi de Rémanences*, trans. B. Vargaftig, B. Zanchi, and J.-B. Para (Sauve, Gard: Clémence Hiver, 1991. Gascar translates this as "leur superbe" [their grandeur].

59 Jean Giono, *Song of the World*, trans. Henri Fluchère and Geoffrey Myers (New York: Counterpoint, 2000), p. 168.

60 Camillo Sbarbaro, *Liquidazione* (Turin: Ribet, 1928).

61 This image of misanthropy and resistance is taken up again by Colombian poet Jorge Cadavid (born in 1962), equally passionate about the natural sciences, in his poem-homage to Camillo Sbarbaro, "Líquenes *(Anthracothecium libricola)*" in *Herbarium* (Bogotá: self-published, 2007).

62 Sbarbaro, *Licheni. Les Lichens* 2 in *Copeaux*, trans. J.-B. Para, pp. 57–66.
63 Giacomo Leopardi, "Le Genêt ou la Fleur du désert." "Broom or the Flower of the Wilderness" in Giacomo Leopardi, *Canti: Poems, A Bilingual Edition*, trans. Jonathan Galassi (New York: Farrar, Straus and Giroux, 2012), pp. 287–289.
64 Sbarbaro, *Licheni. Les Lichens* 4 in *Copeaux*, trans. J.-B. Para, pp. 57–66.
65 Camillo Sbarbaro, *Pianissimo* [Very Softly], (Florence: La Voce, 1914). In *Pianissimo*, trans. B. Vargaftig et al.
66 Sbarbaro, *Fuochi fatui*, in *Copeaux*, trans. J.-B. Para.
67 Sbarbaro, *Pianissimo*, in *Pianissimo*, trans. B. Vargaftig et al.
68 Quote taken from the film by Christophe Bisson on the painter, *Sfumato* (Triptyque Films, 2016, 71 min.).
69 Ibid.
70 Sbarbaro, *Fuochi fatui*, in *Copeaux*, trans. J.-B. Para.
71 Clémence Jeannin demonstrates this superbly in his doctoral thesis entitled, "Bribes et murmures: Étude sur l'aridité et la fragmentation dans l'œuvre en vers et en prose de Camillo Sbarbaro," directed by Yannick Gouchan et defended in November 2017 at the Université d'Aix-Marseille.
72 Sbarbaro, *Trucioli*, in *Copeaux*, trans. J.-B. Para.
73 Sbarbaro, *Pianissimo*, in *Pianissimo*, trans. B. Vargaftig et al.
74 Sbarbaro, *Fuochi fatui*, in *Copeaux*, trans. J.-B. Para.
75 On the work of Pierre Gascar and his relationship to nature, his "ecopoetics," see the works of Pierre Schoentjes, *Ce qui a lieu: Essai d'écopoétique* (Marseille: Wildproject, 2015) and those that he has supervised, like that of Sara Buekens, "La Nature dans l'œuvre de Pierre Gascar: Une étude écopoéticienne" (Master's thesis, Université de Gand, 2015).
76 Pierre Gascar traveled all his life: in Russia, China, Thailand, Indonesia, the Philippines, India, and Ethiopia, as well as in France. *Voyage chez les vivants* (1958) was influenced by a first series of trips to Southeast Asia and Africa. François Sureau wrote : "Humboldt was a traveler and Gascar shares this

penchant for traveling, if not roaming, this dissatisfaction with being there" (*L'Express*, September 24, 1998).

77 Gascar lived for a good part of his life in Lons-le-Saunier, in Jura; his lichen collections are now housed in Besançon (Bibliothèque d'étude et de conservation, Pierre-Gascar Collection, Ms. Z 473).

78 Gascar, *Le Présage*, p. 150.

79 Ibid., p. 148.

80 On the fascination Nerval held for artists in the surrealist period, see Vincent Zonca, "'Je rêvais d'écrire mon *Aurélia*' : l'empreinte nervalienne et l'exploration du désir chez Bernard Noël" in Corinne Bayle (ed.), *Poète cherche modèle* (Rennes: Presses universitaires de Rennes, 2017), pp. 49–62.

81 Pierre Gascar, *Le Fortin* (Paris: Gallimard, 1983).

82 Pierre Gascar, *Portraits et souvenirs* (Paris: Gallimard, 1991): "I cut – a little nervously – the string around the package [from Caillois] and tore off the wrapping: the shoebox was full of lichens."

83 Gascar speaks of the "original plants, almost as old as the planet [...], irreplaceable documents of our evolution," "the most primitive, the most resistant, and the oldest of all flora." As Marc-André Selosse notes in *La Symbiose* (2000) (Paris: Vuibert, 2009), "it is wrong to think [...] that lichens were the first colonizers: photosynthetic micro-organisms already covered the emergent lands in the Precambrian," that is, more than 570 million years ago, 160 million years before the earliest lichen fossil found to date, called *Winfrenatia reticulata*, in Scotland.

84 I am thinking of the excellent critical article by Michel Murat: 'Michel Murat relit *Le Présage* de Pierre Gascar', *Revue critique de fixxion française contemporaine* (2012); on line: revue-critique-de-fixxion-francaise-contemporaine.org/rcffc/article/view/fx11.19/1013.

85 In some of his preparatory notes for the book, Gascar wrote, "I certainly did not need lichens to measure the dangerous effects of radioactive fallout and air pollution, or to discover that the

fight against world hunger, in order not to prove futile, must be based on a new philosophy, nor did I need them to feel how this civilization would suffer from being cut off from the reality of the world" ('Lichens', handwritten manuscript (Bibliothèque d'étude et de conservation de Besançon, Pierre-Gascar Collection, Ms. Z 473).

86 Vincent Zonca, 'Des poétiques de la "décadence" à la fin du XXe siècle? Luis Antonio de Villena et Guy Debord: entre avant-gardes et postmodernités', in Guri Barstad and Karen P. Knutsen (eds.), *States of Decadence: On the Aesthetics of Beauty, Decline and Transgression* (Newcastle upon Tyne, England: Cambridge Scholars Publishing, 2016), vol. 2, pp. 254–268.
87 Gascar, *Le Présage*, p. 29.
88 Ibid., p. 69.
89 Ibid., p. 46.
90 Ibid., p. 24.
91 Pierre Gascar was also very fond of aquariums, which he compared to paintings.
92 Ibid., pp. 150–151.
93 Walter Benjamin, draft notes dating from the preparation of his *Arcades Project*, 'Was ist Aura ?' (1931), French trans. P. Ivernel, in Ursula Marx, Gudrun Schwarz, Michael Schwarz et al., *Walter Benjamin: Archives: Images, textes et signes*, F. Perrier (ed.) (Paris: Klincksieck, 2011), p. 46.
94 Gascar, *Le Présage*, p. 171 [italics mine].
95 Annette Brandt, Jean-Pierre de Vera, Silvano Onofri and Sieglinde Ott, 'Viability of the lichen *Xanthoria elegans* and its symbionts after 18 months of space exposure and simulated Mars conditions on the International Space Station', *International Journal of Astrobiology* (Cambridge: Cambridge University Press, 2014.)
96 The processes of anabiosis were observed in certain microorganisms beginning in the eighteenth century. They were first called "revivicence," then "latent life" (Claude Bernard). The 1860s were also important years for examining this concept.

The word "anabiosis" was coined in 1872 by English scientist William Thierry Preyer.
97 Jean Follain, *Les Uns et les Autres* (Mortemart: Rougerie, 1981.)
98 Joëlle Gardes, *Sous le lichen du temps* (Paris: L'Amandier, 2014).
99 *Lichen*, a Canadian literary journal started in May 1999 in Durham by Rabindranath Maharaj.
100 On this subject, see Marinella Termite, *Le Sentiment végétal: Feuillages d'extrême-contemporain* (Turin: Quodlibet, 2017).
101 Actually, lichens – and mosses – grow on the side of the tree that provides the most moisture; the north is usually more moist because it is protected from the sun, but that can depend on the winds as well.
102 Interview with the author, March 26, 2018.
103 Jaime Siles, *Himnos tardíos*, IV (Madrid: Visor, 1999). "Hymnes tardif 4" in *Hymnes tardifs*, trans. H. Gil (Belval: Circé, 2003).
104 Jaime Siles, *Pasos en la nieve* (Barcelona: Tusquets, 2004).
105 Jaime Siles, *Semáforos, semáforos* (Madrid: Visor, 1990). "Corps diplomatique" in *Sémphores, sémaphores*, trans. H. Gil (Clermont-Ferrand: Presses universitaires Blaise-Pascal, 2013), pp.186–189.
106 "Forman parte de mi simbología y de mi léxico personal," interview with the author, March 26, 2018.
107 Jorge Luis Borges, *El Hacedor* (Buenos Aires: Emecé, 1960). "The Maker," trans. Stephen Kessler, in Alexander Coleman (ed.), *Jorge Luis Borges: Selected Poems* (New York: Viking Penguin, 1999).
108 Jaime Siles, *Música de agua* (Madrid: Visor, 1983). "Métamorphose" in *Musique d'eau, Columnae, Poèmes*, trans. F. Morcillo (Brussels: Le Cri, 1996).
109 These expressions echo the titles of two volumes of collected poems: *Cenotafio: Antología poética (1969–2009)* (Madrid: Visor, 2011); *Cántico de disolución: Poesía (1969–2011)* (Madrid: Verbum, 2015).
110 Henry Gil, *La Poésie de Jaime Siles: Langage, ontologie et esthétique* (Lyon: ENS Éditions, 2014). Henry Gil is the most astute critic of Siles' work, as well as his translator into French.
111 Regarding Jaime Siles' poetry and especially the use of the *silva*

form, see the works of Henry Gil, in particular the article that appeared in *Rhytmica* (2013), in which he discusses the "metrical palimpsest."
112 Interview with the author in November 2017, with Hector Ruiz, scholar of the Golden Age of Spanish literature.
113 Mário Garcia, 'Breve introdução à presença e ao uso da tradição clássica na poesia de Nuno Júdice', Faculty of Letters, University of Coimbra. [Faculdade de Letras da Universidade de Coimbra.]
114 Nuno Júdice, *Lira de Líquen* (Lisbon: Rolim, 1986).
115 Interview with the author, March 26, 2018.
116 Júdice, *Lira de Líquen*.
117 Ibid.
118 In ancient Greek, the verb *leikein* (λείχειν), "to lick," from which comes the word "lichen," is also the root for the word *likanos* which means "the finger one licks," the "index," and by extension gives its name to one of the lyre strings, precisely the one played by the index finger.
119 Nuno Júdice, *Un Canto na Espessura do Tempo* (Lisbon: Quetzal Editores, 1992); *Meditação sobre Ruínas* (Lisbon: Quetzal Editores, 1994). *Un chant dans l'épaisseur du temps, suivi de Méditation sur des ruines*, trans. M. Chandeigne (Paris: Gallimard, 1996), p. 225.
120 Title of a collection by Al Berto, *Salsugem* [Kelp] (Lisbon: Assírio & Alvim, 1984).
121 "Vertical dignity" is an expression used by critic Alain Freixe, evoking Antoine Emaz's poetics.
122 Antoine Emaz, *Os* (Saint-Benoît-du-Sault: Tarabuste, 2004.)
123 Jacques Dupin, "Even If the Mountain 8" in *Jacques Dupin: Selected Poems*, trans. Paul Auster (South Park, Hexham, Northumberland: Bloodaxe Books, 1992).
124 Interview with Antoine Emaz in *Scherzo* (Summer 2001). In the sixteenth and seventeenth centuries, the "ivy poet" was one who embraced "patron trees" with noble roots.
125 Antoine Emaz, *Lichen, encore (notes)* (Paris: Rehauts, 2009).

126 Ibid.
127 Victor Hugo, preface to *Cromwell* (1827), trans. George Ives Burnham, *Cromwell* (Boston: Little, Brown & Co., 1909), p. 164.
128 "¿Por qué se publica esta sencillez, escrita como jugando, y no mis encrespados *Versos libres*, mis endecasílabos hirsutos, nacidos de grandes miedos, o de grandes esperanzas, o de indómito amor de libertad, o de amor doloroso a la hermosura, como riachuelo de oro natural, que va entre arena y aguas turbias y raíces, o como hierro caldeado, que silba y chispea, o como surtidores candentes?" (José Martí, prologue to *Versos sencillos*, 1891). This demand for poetic and formal freedom is expressed through the same metaphor in Martí's poem, "Crin hirsuta." "The Hirsutes" was also the name given to members of a bohemian literary circle in France in the late nineteenth century (1880–1882).
129 Emaz, *Lichen, encore*.
130 Antoine Emaz, *Lichen, lichen* (Paris: Rehauts, 2003).
131 Interview with Antoine Emaz in *Nu* 33 (October 2006) [italics mine].
132 Olvido García Valdés, 'Están muy solos también los animales', interview with Alfonso Armada, *ABC Cultura* (August 14, 2014).
133 Olvido García Valdés, *Del ojo al hueso* (Madrid: Ave del Paraíso, 2001), p. 52.
134 Olvido García Valdés, 'Un poeta se hace con sus enfermedades', *El Ciervo* 636 (March 2004), p. 48.
135 García Valdés, *Del ojo al hueso*.
136 Ibid.
137 Georges Didi-Huberman, *Survival of the Fireflies*, trans. Lia Swope Mitchell (Minneapolis: University of Minnesota Press, 2018), p. 2.
138 Interview with the author, August 2018.
139 García Valdés, *Del ojo al hueso*, p. 55.
140 Interview with the author, August 2018.
141 I recommend this very beautiful work, a personal homage to Lacarrière's writing on the move: Luis Mizón, *Le Sacré bricolage de l'esprit* (Paris: Jean-Michel Place, 2004).

142 Jacques Lacarrière, *À l'orée du pays fertile: Anthologie* (Paris: Seghers, 2011).
143 Jude Stéfan, *Épodes ou Poèmes de la désuétude* (Paris: Gallimard, 1999).
144 This wonder in the face of the perfection of forms in nature, which can only be the work of divine creation, and which serves as model to artists, is a theme in botanical manuals and literary manifestos, at least in those based on religious foundations. Guy Nearing, in *The Lichen Book* (1947), wrote : "But lichens are beautiful too, adorning rocks and trees lavishly with their chaste embroideries. The lobes of a Parmelia may branch as exquisitely as do the ornaments on a marble temple whose design, perhaps, the sculptor borrowed here, from the temples of nature. The fruiting tips of one Cladonia may flaunt a vermilion purer than the garden's most treasured flower, while the goblets of another might serve as models for the acme of the potter's art. These forms stem from the root of creation, primal and pure." *The Lichen Book: Handbook of Lichens of Northeastern United States* (Ashton, MD: Eric Lundberg, 1962), p. 1.
145 Pierre Joseph Van Beneden, *Les Commensaux et les parasites dans le règne animal* (Paris: Baillière, 1875).
146 Crozier, *Small Mechanics*, p. 9.
147 Jacques Lacarrière, *Lapidaires lichens* (Saint-Clément-de-Rivière: Fata Morgana, 1985).
148 Lacarrière, *Lapidaires lichens*.
149 Ibid.
150 On this subject, see for example, Béatrice Bonneville-Humann and Yves Humann (eds.), *L'Inquiétude de l'esprit ou Pourquoi la poésie en temps de crise* (Paris: Éditions nouvelles Cécile Defaut, 2014).
151 Élisée Bec, interview conducted by Pierre Morens and published on line in the review *Infusion* (April 4, 2016).
152 This is Élisée Bec's term, interview with the author, June 3, 2018.
153 "My editorial bias is for short poems, and lichen is a plant that,

although prolific and generous, doesn't take up much room (vertically in any case)," explains Élisée Bec (interview with author, June 3, 2018.

154 Pierre Paolo Pasolini, "Il vuoto del potero in Italia," in *Corriere della sera*, February 1, 1975. This text, also appearing since under the title, "L'article des lucioles," is included in the collection *Écrits corsaires* (1975),
155 Didi-Huberman, *Survival of the Fireflies*, p. 9.
156 Didi-Huberman, *Survival of the Fireflies*, p. 83.
157 On this theme, see Dominique Kalifa, *Les Bas-fonds: Histoire d'un imaginaire* (Paris: Seuil, 2013).
158 Let us recall the regional and popular dialects, more politicized of course, in the work of Pasolini, and what Didi-Huberman says of them in luminous terms in his *Survival of the Fireflies*, pp. 11–12.
159 Marcel Schwob, *Étude sur l'argot français* (1889) (Paris: Allia, 2004), p. 59 ; Yves Chaudouët admires this author.
160 Peter Stephan Jungk, "*Auf der Suche nach dem Irrlicht*" ["In Search of the Will-o'-the-Wisp"], preface to the catalogue for an exhibition at the Montenay-Giroux Gallery, Paris, March 1996.
161 Butor, *Transit A*. [italics mine].
162 Georges Duhamel, 'L'Insurrection des humbles', in *Le Bestiaire et l'Herbier* (Paris: Mercure de France, 1948).
163 Zola, *Abbé Mouret's Transgression*, trans. E.A. Vizetelly, pp. 251–252.
164 Elizabeth Bishop, *Questions of Travel* (New York: Farrar, Straus & Giroux, 1968).
165 Brenda Hillman, *Extra Hidden Life, Among the Days* (Middletown: Wesleyan University Press, 2018).
166 Bachelard, *The Poetics of Space*, p. 181.
167 We can see this, for example, in the work of Jacques Réda (*Les Ruines de Paris*, 1977; *L'Herbe des talus*, 1984; *Beauté suburbaine*, 1985), Jacques Ancet (*Lisières*, 1985), the Portuguese writer Nuno Júdice (*Canto in the Thickness of Time*, 1992; *Meditation*

on Ruins, 1994 – with poems such as "Suburban Landscape," "Decadence," "Fallow Land"), Jean Rolin ("Zones," 1995), Dominique Fourcade, Bernard Noël, Emmanuel Hocquard (*Ruines à rebours*, 2010), Pierre Alferi, Jude Stéfan (and his "*prosèmes de grenier*"), Yann Miralles (*Des terrains vagues, variations*, 2016), Jean-Marie Gleize (*Trouver ici: Reliques et lisières*, 2018). As well as in reviews, such as *Décharge* (created in 1981) and *Friches* (created in 1983).

168 Jean-Michel Maulpoix, preliminary notes written in December 2000 for his *Pour un lyrisme critique* (Paris: José Corti, 2009), source : author's website.
169 Philippe Vasset, *Un livre blanc* (Paris: Fayard, 2007), p. 100.
170 Marc Jeanson and Charlotte Fauve, *Botaniste* (Paris: Grasset, 2019), p. 175.
171 Audrey Muratet, Myr Muratet and Marie Pellaton, *Flore des friches urbaines du nord de la France et des régions voisines* (Paris: Xavier Barral, 2017), p. 7.
172 Gilles Clément, *Manifeste du tiers-paysage*, (Paris: Éditions Sujet/ Objet, 2004); on line: www.gillesclement.com/cat-tierspaysage-tit-le-Tiers-Paysage.
173 Ibid.
174 See on this subject Cardim's website: www.cardimpaisagismo.com.br/projetos/floresta-de-bolso.
175 Gilles Clément, *Éloge des vagabondes* (Paris: Robert Laffont, 2014), p. 11.
176 Oscar Furbacken, artist statement on his website: www.oscarfurbacken.se expeditionbeyond.html.
177 Oscar Furbacken, artist statement on his website: www.oscarfurbacken.se/expeditionbeyond.html.
178 Friedensreich Hundertwasser, "Mouldiness Manifesto against Rationalism in Architecture" (1958/1959/1964): https://www.hundertwasser.at/english/texts/philo_verschimmelungsmanifest.php.

4 Toward a Symbiotic Way of Thought

1 Symbiosis was conceived in the nineteenth century essentially beginning with two models coming from fungi: lichenized fungi and mychorrhizea .
2 Selosse, *La Symbiose*, p. 5.
3 Ibid., p. 134.
4 Stephen Collis, Canadian poet, activist, and academic, reflects on this idea in his writings, notably in *To the Barricades* (Vancouver: Talonbooks, 2013).
5 *Datableed* 3 (2015) (https://www.datableedzine.com/drew-milne-lichens), then in, Drew Milne, *In Darkest Capital: Collected Poems* (Manchester: Carcanet Press Ltd (2017), section "Lichens for Marxists."
6 Karl Friedrich Wilhelm Wallroth, *Naturgeschichte der Flechten* (Frankfurt am Main: Bey Friedrich Wilmans, 1825–1827).
7 Simon Schwendener, 'Untersuchungenüberden Flechtenthallus', in *Beiträge zur wissenschaftliche Botanik* 4 (1868), pp. 195–207.
8 Albert Bernhard Frank, *Über die biologischen Verhältnisse des Thallus einiger Krustenflechten, Beiträge zur Biologie der Pflanzen* [On the biological conditions of the thallus of certain crustose lichens] (1877), vol. 2, pp. 123–200 [italics mine].
9 This definition of symbiosis is found in *Die Erscheinung der Symbiose* [*The Appearance of Symbiosis*], translated into French in 1879. Oliver Perru and Marc-André Selosse stress that Anton de Bary no doubt adopted Albert Bernhard Frank's term, whom he knew and whose works he admired, as their universities were in the same vicinity (contrary to a version of history that makes Anton de Bary the inventor of the word). Whatever the case, the development of this concept was concomitant and clearly shared by these two researchers.
10 Ernst Haeckel, *Generelle Morphologie der Organismen* [*General Morphology of Organisms*] (Berlin: Georg Reimer, 1866).
11 Chantal Van Haluwyn, 'La symbiose lichénique: un partenariat

difficile à démasquer', *Bulletin de l'Association française de lichénologie* 45:2 (November 2020), p. 158.

12 David Leslie Hawksworth and Martin Grube, 'Lichens redefined as complex ecosystems', *New Phytologist* 227:5 (2020), pp. 1281–1283.

13 "Regarding the term itself, it was used for the first time and in a political sense by the Greek historian Polybius, in the second century BCE." (Olivier Perru, 'Aux origines des recherches sur la symbiose vers 1868–1883', *Revue d'histoire des sciences* 59: 1 (2006), p. 18, note 52).

14 The word "cryptogam," used to designate the family to which lichens belong, means in ancient Greek, "hidden marriage," "hidden birth" (because their reproduction is invisible to the naked eye).

15 I have found no evidence of a single scientist who mentions this.

16 This is in Volume 14 of his *Moral Essays.*

17 [Italics mine]. In ancient Greek: "ἀλλὰ μᾶλλον ἐπὶ τὰς κατ' ἰδίαν κοινωνίας αὐτῶν καὶ συμβιώσεις ἰτέον."

18 See especially an extract from Chrysippus' *Ten Treatises on the Honorable and on Pleasure*, cited by Athenaeus of Naucratis (*Deipnosophistea*, 3, 38, 89d).

19 Plutarch, *De Sollertia Animalium* 30, trans. William W. Goodwin et al., in *Plutarch's Morals* (Boston: Little, Brown and Company, 1878), vol. 5 [italics mine].

20 In 'Apologie de Raymond Sebond' [An Apology for Raymond Sebond], (*Essays* 2, 12), very much inspired by Plutarch, Montaigne mentions various examples of symbiosis: those "particular offices that we draw from one another, for the service of life."

21 Ambroise Paré, *Opera Ambrosii Parei* 24, 'De monstris et prodigiis' (Paris: Jacob du Puys Éditeur, 1582), p. 781 : "Chrysippus Solensis in *de Honest. & volup.*: Pinna & pinnoter operas *mutuas* praestat, seorsum vivere non possunt" [italics mine]; this case is mentioned just after that of the hermit crab.

22 Theophratus, *Enquiry into Plants and Minor Works on Odours and Weather Signs*, trans. Sir Arthur Hort (Loeb Classical

Libarary, London: William Heinemann, New York: G.P. Putnam and Sons, 1946), vol. 5, 2, 2, p. 105 [italics mine]. This is no doubt a reference to grafting, which we also find in Pliny (*Natural History*, Book 17, ch. 24, 101). Speaking of the discovery of grafting [*coitum*], he writes: "A careful husbandman, being desirous, for its better protection, to surround his cottage with a palisade, thrust the stakes into growing ivy in order to prevent them from rotting. Seized by the tenacious grasp of the still living ivy, the stakes borrowed life from the life of another wood, and it was found that the stock of a tree acted in place of earth." (trans. John Bostock, Henry T. Riley [London: Henry G. Bohn, 1856]).

On line: http://www.perseus.tufts.edu/hopper/text?doc=Perseus:text 1999.02.0137:book=17:chapter=24&highlight=ivy

23 The revolution in scientific knowledge brought about by revolutionary nineteenth-century technology cannot be emphasized enough. Those advances allowed for the discovery of lichenic symbiosis as well as an understanding of how certain diseases work and the implementation of vaccines, for example. Currently, we are witnessing a new revolution, involving the study of lichens, thanks to progress in genetic analysis.

24 On this subject, see the very beautiful edition of an extract from the *Théorie de l'unité universelle*, entitled *La Botanique passionnelle* (Vichy: La Brèche, 2017).

25 Charles Fourier, *Traité de l'association domestique-agricole* (Paris and London, 1822), p. 499.

26 Van Beneden, *Les Commensaux et les parasites dans le règne animal*, p. 3.

27 Pierre-Joseph Proudhon, *Théorie de la propriété* (posth.) (Paris: Flammarion, 1871) [italics mine].

28 Van Beneden, *Les Commensaux et les parasites dans le règne animal*, pp. 10–11 [italics mine].

29 "It is interesting to see that J.-P. Van Beneden, professor in Leuven, based his whole approach to interrelationships in the animal kingdom [...] on analogies with human societies." Olivier

Perru, *De la société à la symbiose: Une histoire des découvertes sur les associations chez les êtres vivants [1860–1930]* (Paris: Vrin, 2003).

30 "Biological mutualism is thus in part the avatar of philosophical and political thinking that believes in the solidarity of altruism, at the risk of anthropomorphism and finalism when it is projected onto biology. [...] The concept of biological mutualism emerged in the era when workers unionized into "mutualities" for mutual protection from the hazards of existence." Marc-André Selosse, 'Symbiose et mutualisme *versus* évolution: De la guerre à la paix?', *ATALA*, "Pour une biologie évolutive," 15 (2012).

31 The thesis that biological mutualism is rooted in the political ideology of the period was developed especially in the 1980s and 1990s by Douglas H. Boucher (*The Biology of Mutualism: Ecology and Evolution* (New York: Oxford University Press, 1985), pp. 13–14) and by Jean-Marc Drouin (*L'Écologie et son histoire* (Paris: Flammarion, 1993)), to locate them in this specific context. In his thesis (*Biologie et complexité: Histoire et modèles du commensalisme*, directed by Olivier Perru, University of Lyon- I (2015), pp. 231–236), Brice Poreau adds nuance to the question of how directly the political influenced the biological in the thinking on mutualism, with evidence from his meticulous study of Van Beneden's archives. Poreau focuses particularly on this scientist, while simultaneously demonstrating the importance of the concept in the history of biology.

32 See on this subject, Daniel Todes, 'Global Darwin: Contempt for competition', *Nature* 462 (2009), pp. 36–37; Marc-André Selosse, 'Symbiose et mutualisme *versus* évolution: De la guerre à la paix?', article cited above, note 30.

33 The works of Lyon researcher Olivier Perru are illuminating with regard to the archeology of associative thinking on living beings beginning in the 1860s, the reinterpretation of the theory of evolution, and the contribution of genetics: *De la société à la symbiose: Une histoire des découvertes sur les associations chez les êtres vivants [1860–1930]*, 2 vol. (see note 29 above).

34 Karl Fedorovitch Kessler, "On the Law of Mutual Aid," speech given in December 1879.
35 Pierre Kropotkine, *Mutual Aid: A Factor in Evolution* (London: William Heinemann, 1902).
36 See on this subject, the works of French botanist and symbiosis specialist Marc-André Selosse, *Jamais seul: Ces microbes qui construisent les plantes, les animaux et les civilisations* (Arles: Actes Sud, 2017); and of French researcher in evolutionary biology Éric Bapteste, *Tous entrelacés: Des gènes aux super-organismes : les réseaux de l'évolution* (Paris: Belin, 2018).
37 Let us think, for example, of the book by Pablo Servigne and Gauthier Chapelle, *L'Entraide: L'autre loi de la jungle* (Paris: Les Liens qui libèrent, 2017). *Mutual Aid: The Other Law of the Jungle*, trans. Andrew Brown (Cambridge: Polity Books, 2022).
38 See Francis Hallé, *Éloge de la plante: Pour une nouvelle biologie* (Paris: Seuil, 1999), and *Plaidoyer pour l'arbre* (Arles: Actes Sud, 2005).
39 Patrick Blanc, *Être plante à l'ombre des forêts tropicales* (Paris: Nathan, 2002).
40 There are, among others, the outstanding works by David George Haskell (*The Forest Unseen: A Year's Watch in Nature*, 2012), Stefano Mancuso (*The Revolutionary Genius of Plants*, 2013), Michael Marder (*Plant-Thinking: A Philosophy of Vegetal Life*, 2013), Eduardo Kohn (*How Forests Think: Toward an Anthropology Beyond the Human*, 2013), Anthony Trewavas (*Plant Behavior and Intelligence*, 2014), Peter Wohllenben on the subject of mycorrhizae (*The Hidden Life of Trees*, 2015, a best-seller exploring the networks of cooperation and communication of root systems in forests), Anna Lowenhaupt Tsing (*The Mushroom at the End of the World: On the Possibility of Life in Capitalist Ruins*, 2015), Natasha Myers (who proposed the idea of a "phytocene age" in 2016, and, evoking photosynthesis, emphasized that it is plants that "make the world"), Jacques Tassin (*What do Plants Think About?*, 2016), Emanuele Coccia (*The Life of Plants: A Metaphysics of Mixture*, 2016), Marc-André Selosse

(*Jamais seul: Ces microbes qui construisent les plantes, les animaux et les civilisations*, 2017), Jean-Baptiste Vidalou (*Être forêts*, 2017), and Eric Bapteste (*Tous entrelacés: Des gènes aux super-organismes: les réseaux de l'évolution*, 2018). Listed here are the publication dates of the texts in translation or in their original languages. Let us note that in October 2020, The Fungi Film Festival, the first international festival of short films devoted to fungi, was held on a mushroom farm in Portland, Oregon in the United States, and viewed on line internationally.

41 Let us mention *Gathering Moss: A Natural and Cultural History of Mosses* (2003) by Robin Wall Kimmerer of the United States, *Mosses, My Dear Friends* (2011) by Hisako Fujii of Japan, *Louange des mousses* (2012) by Véronique Brindeau of France, and *Moss: From Forest to Garden: A Guide to the Hidden World of Moss* (2016) by Ulrica Nordström of Sweden.

42 Lynn Margulis, 'Symbiogenesis and symbioticism', in Lynn Margulis and René Fester (eds.), *Symbiosis as a source of evolutionary innovation* (Cambridge: MIT Press, 1991), pp. 1–14 (p. 4 for the citation).

43 Paul Nardon, 'Rôle de la symbiose dans l'adaptation et la spéciation', *Bulletin de la Société zoologique de France* 70:4 (1995), pp. 397–406.

44 Olivier Perru, 'Aux origines des recherches sur la symbiose vers 1868–1883', *Revue d'histoire des sciences* 59:1 (2006), p. 18, note 52.

45 David C. Smith, 'Symbiosis research at the end of the millennium', *Hydrobiologia* 461 (2001), pp. 49–54.

46 Julio Cortázar, *Hopscotch*, trans. Gregory Rabassa (New York: Random House, 1966), p. 236.

47 Haskell, *The Forest Unseen*, p. 228.

48 Ibid., p. 3.

49 Emanuele Coccia, *Metamorphosis*, trans. Robin Mackay (Cambridge: Polity, 2021), p. 3.

50 Haskell, *The Forest Unseen*, p. 5.

51 S.F. Gilbert, Jan Sapp, A.I. Tauber, 'A symbiotic view of life: We

have never been individuals', *Quarterly Review of Biology* 87 (2012), pp. 325–341.
52 Olga Potot, "Nous sommes tou·te·s du lichen' : Histoires féministes d'infections trans-espèces', *Chimères* 82:1 (2014), pp. 137–144.
53 Karine Prévot, 'Sommes-nous des lichens? Une perspective végétale sur l'individu', *Critique* 850: 3 (2018), pp. 204–213.
54 Donna J. Haraway, *Staying With the Trouble : Making Kin in the Chthulucen* (Durham: Duke University Press, 2016), p. 72.
55 Since January 2021, fungi and lichens are even more present in the artistic world and this slogan even more fashionable. Of course this sometimes indicates an idealization of lichenic symbiosis, a kind of perfect mutualism appearing as a new ethical horizon. I have come upon it in exhibitions, like at the Brazilian Museum of Sculpture and Ecology in São Paulo in a series of flags done in 2020–2021 by Joana Amador and Mariana Lacerda. Entitled *Little Demonstration*, they offer variations on the theme of struggle ("Struggle like lichen," "Struggle like coral," "Struggle like a refugee," and so on) and point the artistic installation toward a street demonstration, in line with Laura C. Carlson. In the Kirchner Cultural Center it appears in the title of a sculpture by Romina Orazi (born in 1972): *Mundos sobre rastros, todos somos líquenes* (2021) represents a human being in cement covered with fruticose lichens and holding a plant.
56 Artist statement on the website : artpark21.org/.
57 Karine Prévot, 'Sommes-nous des lichens? Une perspective végétale sur l'individu', p. 211, article cited above note 53.
58 Marc-André Selosse, *Jamais seul: Ces microbes qui construisent les plantes, les animaux et les civilisations* (Arles: Actes Sud, 2017).
59 *Voyage avec Michel Butor*, interviews with M. Santchi, p. 185 [italics mine].
60 Michel Butor, *Illustrations* (Paris: Gallimard, 1976).
61 Gilbert Simondon, *Individuation in light of notions of form and information*, trans. Taylor Adkings (Minneapolis: University of Minnesota Press, 2020), p. 201.

62 Ibid., pp. 199–200.
63 F. Dal Grande, G. Rolshausen, P.K. Divakar, A. Crespo, J. Otte, M. Schleuning and I. Schmitt, 'Environment and host identity structure communities of green algal symbionts in lichens', *New Phytologist* 217:1 (2017), pp. 277–289.
64 Goward, 'Twelve Readings on the Lichen Thallus', article cited above, book 1, note 37.
65 The Salish, Samí, Sherpa, and Okanagan vernacular names bear precise traces of a specific relationship and knowledge regarding lichen taken as a whole (see on this subject, Catherine Kendig "Ontology and values anchor indigenous and grey nomenclatures: a case study in lichen naming practices among the Samí, Sherpa, Scots, and Okanagan," *Studies in History and Philosophy of Science Part C: Studies in History and Philosophy of Biological and Biomedical Sciences* 84 (December 2020).
66 Arthur George Tansley, 'The use and abuse of vegetational concepts and termes', *Ecology*, 16:3 (1935), pp. 284–307.
67 Margulis and Fester, *Symbiosis as a source of evolutionary innovation*.
68 Forest Rohwer et al., 'Diversity and distribution of coral-associated bacteria', *Marine Ecology Progress Series* 243 (2002), pp. 1–10.
69 Marc-André Selosse, 'Au-delà de l'organisme, l'holobionte', *Pour la science* 469 (November 2016), pp. 81–84.
70 Gilles Deleuze and Félix Guattari, *A Thousand Plateaus: Capitalism and Schizophrenia*, trans. Brian Massumi (Minneapolis: University of Minnesota Press, 1987), p. 321.
71 Dénètem Touam Bona, 'Lignes de fuite du marronnage: Le "lyannaj" ou l'esprit de la forêt', *Multitudes* 70:1 (2018), pp. 177–185.
72 See on this subject, Marc-André Selosse, 'Les végétaux existent-ils encore?', *Pour la science: Dossier* 77 (October 5, 2012); on line: www.pourlascience.fr/sd/botanique/les-vegetaux-existent-ils-encore-6987.php.
73 In this regard, I'm thinking of the work of Philippe Descola

and the book by Eduardo Kohn, *How Forests Think: Toward an Anthropology beyond the Human* (Berkeley: University of California Press, 2013).
74 Coccia, *Metamorphoses*, trans. Robin Mackay, p. 158–159.
75 Sylviane Carin, 'Expo à Dignac: Pascale Gadon à la croisée de l'art et de la biologie', *Charente libre* (November 28, 2013).
76 On line: www.lichen.fr/perso_adresse.htm.
77 Interview with the author, April 6, 2018.
78 Ibid.
79 'Two Ways of Knowing: Robin Wall Kimmerer on Scientific and Native American Views of the Natural World', interview with Robin Wall Kimmerer by Leath Tonino, *The Sun Magazine* 484 (April 2016), pp. 4–14; on line: https://www.thesunmagazine.org/issues/484/two-ways-of-knowing.
80 Michel Serres, *Le Contrat naturel* (1990) (Paris: Le Pommier, 2018), p. 67. *The Natural Contract*, trans. Elizabeth MacArthur and William Paulson (Ann Arbor: University of Michigan Press, 1995), p. 38.
81 Ibid.
82 Enzensberger, *Blindenschrift*, extract from the poem, "Lichenology."

Envoi: Sporules

1 Haraway, *Staying With the Trouble*, p. 56.
2 Marina Julienne, 'Joël Boustie, rare explorateur du monde des lichens', *Le Monde* (January 4, 2020).
3 Timothy M. Lenton, Tais W. Dahl, Stuart J. Daines, Benjamin J.W. Mills, Kazumi Ozaki, Matthew R. Saltzman, and Philipp Porada, 'Earliest land plants created modern levels of atmospheric oxygen', *Proceedings of the National Academy of Sciences* 113: 35 (August 30, 2016), pp. 9704–9709.

Index of Names

Acharius, Erik, 15, 100
Acloque, Alexandre, 28–9
Akhmatova, Anna, 114, 150
Al Berto, 142
Albert the Great, 4
Altdorfer, Albrecht, 58
Althusius, Johannes, 183
Amoreux, Pierre Joseph, 26
Annunzio, Gabriele D', 113, 124
Antisthenes, 181
Apollinaire, Guillaume, 162, 176
Arcimboldo, Giuseppe, 40, 71
Aristotle, 180
Ascunción Silva, José, 176

Babstock, Ken, 4, 134, 147
Bachelard, Gaston, 1, 39, 61, 63, 96, 161, 219
Bacon, Joséphine, 22, 134
Balzac, Honoré de, 59
Bancquart, Marie-Claire, 134, 159
Baranetzky, Josep Wasilijevitsch, 177
Battistelli, Leo, 43, 58, 131–2, 168, Ill. 8–9
Baudelaire, Charles, 26–7, 58, 113, 121, 139, 145, 178, 185, 204, 212
Bec, Élisée, 155–6
Benjamin, Walter, 130, 166, 168, 212
Bernaer, Richard, 49
Bishop, Elizabeth, 160

Blanc, Patrick, 189
Bonnet, Charles, 109
Borges, Jorge Luis, 137
Boulez, Pierre, 68
Boustie, Joël, 217
Bouvier, Nicolas, 99, 151
Bramsen, Christian, 65
Breton, André, 53–4
Brindeau, Véronique, 77, 78
Brisson, Théodore-Polycarpe, 133
Brocq, Louis, 8
Butor, Michel, 67, 68, 69–71, 86, 159, 196–7, 218

Cage, John, 3, 65, 68, 72–6, 115, 126
Caillois, Roger, 11, 52, 53, 55, 67, 69, 82, 126, 153, 185
Calmettes, Tiphaine, 115
Calvino, Eva Mameli, 116
Calvino, Italo, 116
Candolle, Augustin Pyrame de, 93
Cardim, Ricardo, 165–6, 167
Carlson, Laura C., 58, 60, 194, 208
Caullery, Maurice, 187
Cendrars, Blaise, 176
Césaire, Aimé, 153
Char, René, 150
Chaudouët, Yves, 58, 64–6, 157–8, Ill. 13
Chen Hongshou, 86, Ill. 7
Chokuo, Tamara, 86

Index of Names

Chrysippus, 182
Clément, Gilles, 164–6
Clerc, Philippe, 12–13, 16, 173
Coccia, Emanuele, 5, 37, 205–6, 210
Collis, Stephen, 176
Cortázar, Julio, 191
Courbet, Gustave, 39, 40
Crozier, Lorna, 9, 134, 153

da Vinci, Leonardo, 50, 54
Dalí, Salvador, 40, 71, 72
Dante, 118–19, 130
Darío, Rubén, 138
Darwin, Erasmus, 34, 59, 106–7
de Bary, Anton, 15, 177, 179–80, 181, 190
de Chirico, Giorgio, 40, 71
Deleuze, Gilles, 203
Didi-Huberman, Georges, 81–2, 118–19, 130, 150, 156–7
Dillenius, Johann Jacob, 14–15
Donne, John, 176
Dotremont, Christian, 57–8
Du Bouchet, André, 144
Duchamp, Marcel, 40
Dufau, Roland, 65
Duhamel, Georges, 159
Dupin, Jacques, 134, 143
Durdilly, Hélène, 146

Eliasson, Olafur, 189
Eliot, T.S., 162
Emaz, Antoine, 115, 123, 124, 134–5, 142–8, 155, 169
Enzensberger, Hans Magnus, 32–3, 56–7, 91, 134, 213
Ernst, Max, 53
Escher, M.C., 58, 71, 72
Ettmüller, Michael, 27–8

Famintsyne, Andreï Sergueïevitch, 177, 178
Fauve, Charlotte, 163–4
Fenollosa, Ernest, 87
Flaubert, Gustave, 114, 126
Follain, Jean, 133, 134, 145
Fontaine, Natasha Kanapé, 23–4
Fouque, Thomas, 58, 61
Fourier, Charles, 185–6
Frank, Albert Bernhard, 179–80, 181
Freitag, Gérard, 135
Frénaud, André, 147
Freud, Sigmund, 175, 195
Friedrich, Casper David, 142
Fujii, Hisako, 78
Furbacken, Oscar, 45, 61–3, 104, 130, 167, 168, 170–1, Ill. 10–11

Gadon-González, Pascale, 29, 184, 193, 206–9, Ill. 14–16
García Valdés, Olvido, 134, 148–51
Gardes, Joëlle, 134
Gascar, Pierre, 7, 13, 18, 21, 24, 35, 42–3, 58, 62, 66–7, 98, 109, 124–31, 132, 144, 156, 157, 169, 218
Gatti, Armand, 67, 68, 69
Gaugin, Paul, 137
Gavard-Perret, Jean-Paul, 41
Gette, Paul-Armand, 41
Giacometti, Alberto, 123, 142, 146
Gide, André, 162
Gil, Henry, 138
Gilbert, Scott Frederick, 194
Giono, Jean, 63, 119–20, 147
Goethe, Johann Wolfgang von, 107, 113
Góngora, Luis de, 139
Goward, Trevor, 26, 201, 208

Grube, Martin, 180
Guattari, Félix, 203
Guillevic, Eugène, 123, 134, 145

Haeckel, Ernst, 58, 180
Halle, Francis, 189
Hang, Ren, 42
Haraway, Donna, 192, 194, 216–17
Haskell, David George, 2, 80, 191–2, 194
Hawksworth, David Leslie, 181
Hecataeus, 181
Helder, Herberto, 118, 142
Hillman, Brenda, 34, 56, 134, 147, 160–1
Hiroshige, Utagawa, 86
Hiroshige II, 24
Hoffman, Georg Franz, 26
Hōitsu, Sakai, 87
Hugo, Victor, 1, 10, 112–13, 145, 146
Hundertwasser, Friedensreich, 168
Hutchinson, Peter, 115
Huysmans, Joris-Karl, 9, 10, 49, 162

Ivernois, Jean-Antoine d', 92

Jaccottet, Philippe, 85, 134
Jeanson, Marc, 163–4
Johns, Jasper, 75
Júdice, Nuno, 134, 139–42
Jungk, Peter Stephan, 158

Kahlo, Frida, 40
Kant, Immanuel, 175
Kawabata, Yasunari, 82
Kessler, Karl Fedorovitch, 187–8
Kimmerer, Robin Wall, 35, 211
Kircher, Athanasius, 53
Klimt, Gustav, 87

Kobayashi, Yutaka, 110
Kohn, Eduardo, 210–11
Kōrin, Ogata, 87, 88, 89
Kropotkine, Pierre, 188

Lacarrière, Jacques, 134, 135, 146, 151–5
Lamarck, Jean-Baptiste de, 5, 57
Legay, Bernard, 42, 122–3
Leibniz, Gottfried, 175
Leibowitz, René, 68
Leopardi, Giacomo, 117, 121
Leroi-Gourhan, André, 79–80
Lestel, Dominique, 211
Levinas, Emmanuel, 5
Li Cheng, 85
Ligeti, György, 76
Linnaeus, Carl, 14–15, 100, 106
Long, Lois, 74
Lopez, Barry, 50, 122
Losi, Claudia, 58, 59–60
Lovelock, James, 188
Lucretius, 175
Luxemburg, Rosa, 161

Malesherbes, Chrétien-Guillaume de Lamoignon de, 3, 93
Margulis, Lynn, 188–90, 202
Marti, José, 146
Marx, Karl, 178, 186
Masson, André, 67
Maulpoix, Jean-Michel, 162
Medrano, Espinosa, 139
Melville, Herman, 114
Merejkovski, Constantin Sergeïevitch, 188
Meskache, Djamel, 146
Messagier, Matthieu, 33, 126
Michaux, Henri, 57, 68, 91, 123, 195

Michon, Pierre, 147
Milne, Drew, 158, 176
Monod, Théodore, 19
Montaigne, Michel de, 146, 165, 182, 195
Montale, Eugenio, 117
Munro, Alice, 38–41
Muratet, Audrey, 164
Muratet, Myr, 164
Myers, Natasha, 211

Nardon, Paul, 190, 202
Närhinen, Tuula, 42
Nearing, Guy, 3, 52, 60, 73, 74
Néraud, Jules, 3, 100
Neruda, Pablo, 55, 59, 117
Nerval, Gérard de, 125–6
Ni Zan, 85
Nietzsche, Friedrich, 119, 175
Nilsson, Nagnus, 22
Noël, Bernard, 156
Nylander, Wilhelm, 20, 100, 108–9, 178, 180

Olson, Charles, 72
Osborne, Gillian Kidd, 114
Ovid, 38, 40

Paracelsus, 27, 29
Paré, Ambroise, 182–3
Parks, Rosa, 161
Pasolini, Pierre Paolo, 156
Paz, Octavio, 139
Pellaton, Marie, 164
Perec, Georges, 3, 95–6, 144
Perru, Olivier, 187, 190
Philo of Alexandria, 182
Piero di Cosimo, 54
Pinson, Jean-Claude, 134
Pitxot, Antoni, 58, 71–2, Ill. 5

Plato, 183
Pliny the Elder, 29–30, 31
Plutarch, 182
Polybius, 181
Ponge, Francis, 2, 41, 118
Poreau, Brice, 187
Potot, Olga, 194
Poussin, Nicolas, 68
Prévot, Karine, 194
Proudhon, Pierre-Joseph, 186
Proust, Marcel, 2, 11

Qianlong, 4
Quevedo, Francisco de, 138

Rabelais, François, 14
Rauschenberg, Robert, 72, 73, 75, 115
Reverdy, Pierre, 145
Rimbaud, Arthur, 113–14, 125–6, 195
Robert, Hubert, 40
Robinson, William, 165
Rohwer, Forest, 202
Rolin, Jean, 4
Roshū, Fukae, 87
Ross, Dieter, 115
Rousseau, Jean-Jacques, 3, 62, 70, 92–100, 116, 117, 121, 145, 147–8, 151, 152, 212
Ruskin, John, 60, 99

Saby, Bernard, 58, 67–70, 72, 78, 125, 196
Sand, George, 3, 52, 53, 100–1
Sanraku, Kanō, 86
Sansetsu, Kanō, 86
Saraceno, Tomás, 189
Saussure, Horace Bénédict de, 93
Sbarbaro, Camillo, 20, 36, 45,

50–1, 57, 59, 66–7, 115, 116–24, 131, 132, 144, 145, 152, 169, 208, 218
Schwendener, Simon, 177–9, 191, 199
Schwob, Marcel, 157
Second, Claire, 50
Segalen, Victor, 153
Seghers, Hercule, 58
Selosse, Marc-André, 175, 187, 195, 202–3
Serres, Michel, 57, 173, 212
Seurat, Georges, 68
Shakespeare, William, 7–8, 71
Siles, Jaime, 33, 80, 134, 136–9, 152
Simondon, Gilbert, 199–200
Smith, Alexander Hanchett, 74
Smith, David Cecil, 190
Sonfist, Alan, 165–6
Soseki, Musō, 78
Sosen, Mori, 87
Starobinski, Jean, 165
Stéfan, Jude, 134, 152
Suzuki, Daisetz Teitaro, 73
Suzuki, Kiitsu, 87

Taigong Wang, 88
Tanizaki, Jun'ichirō, 81
Tansley, Arthur George, 201–2
Tesson, Sylvain, 6
Theophrastus, 4, 13, 31, 183
Thoreau, Henry David, 2, 51–2, 67, 73, 102–6, 114, 151, 152, 161, 169, 180, 189, 218, 219
Touam Bona, Dénètem, 203
Tournefort, Joseph Pitton de, 14
Tournier, Michel, 38
Tranströmer, Tomas, 55, 134
Tsing, Anna L., 217
Tudor, David, 72
Tujague, Mathias, 60–1

Ungaretti, Giuseppe, 117

Vainio, Edvard, 100
Valéry, Paul, 93
Van Beneden, Pierre-Joseph, 152–3, 179, 183, 185, 186
Vasari, Giorgio, 54
Vasset, Philippe, 163
Veyrat, Marc, 22
Voltaire, 94

Wallroth, Karl Friedrich Wilhelm, 177
Wang Qian, 86
Wang Tingyun, 85
Wasselin, Lucien, 33, 135
Whitman, Walt, 113, 119
Willemet, Pierre Rémi, 26
Willink, Albert Carel, 142
Wordsworth, William, 107
Wyndham, John, 134

Xenakis, Iannis, 76–7, 146

Yu, Kouo, 27

Zao Wou-Ki, 68
Zerbini, Luiz, 209–10, Ill. 12
Zola, Émile, 7, 9, 113, 159–60, 218

Index of Lichens

Acarosporaceae (Gn.), 215
Alectoria (Gn.), 110

Bryoria (Gn.), 110
Bryoria fremontii (or Horsehair lichen), 25–6
Buellia frigada, 32

Caloplaca (Gn.), 18, 49
Caloplaca aurantia, 49
Candelariella vitellina, 49, 161
Cetraria islandica (or Iceland lichen), 24, 26
Cladonia (Gn.), 48
Cladonia arbuscula, 46
Cladonia foliacea, 20
Cladonia portentosa, 46
Cladonia stellaris (or Reindeer lichen), 22, 57
Cladonia uncialis, 10
Dictyonema huaorani (or Nënëndapë), 24
Diploschistes scruposus, 66

Evernia prunastri (or Oak moss), 30

Flavopunctelia soredica, 56

Graphidaceae (Gn.), 56, Ill. 4
Graphis scripta (or Script lichen/Secret writing lichen), 56, 57, 154

Herpothallon rubrocintum (or Christmas lichen), 49, 110, Ill. 2
Hypotrachyna laevigata, 51

Icmadophilia ericetorum, 10

Lecanora carpinea, 46
Leconora esculenta (or Manna of the desert), 4, 24
Lecanora miniatus, 46–7
Lecanora muralis, 220
Lecanora parella, 30
Lepraria (or Leprose) (Gn.), 9, 37
Lobaria pulmonaria (or Lungwort lichen), 27, 104

Parmelia (Gn.), 24, 32
Parmelia vagans, 36
Parmelia saxatilis, 27
Parmelia sulcata, 16
Parmotrema (Gn.), 86–7
Parmotrema tinctorum, 86
Peltigera aphthosa (or Green dog lichen), 26, 27
Pertusaria pertusa, 46
Phaeographina (Gn.), Ill. 4
Phaeophyscia hirsuta, 10
Physcia (Gn.), 205
   Rhizocarpon geographicum (or Map lichen), 32, 63, 132–3

Index of Lichens

Usnea barbata (or Beard lichen, barba de viejo), 10, 26, 27
Pleurosticta acetarbulum, 32
Pseudevernia furfuracea (or Tree moss), 30, 177
Pygmaea (Gn.), 18

Ramalina (Gn.), 133
Ramalina menziesii (or Lace lichen), 56, 58–9
Rhizocarpon geographicum (or Geographic lichen), 32, 63, 132–3
Roccella (or Orchilla) (Gn.), 32

Umbilicaria (Gn.), 35, 37
Umbilicaria esculenta (or Rock tripe/*Iwatake*), 24

Usnea (Gn.), 14, 18, 20, 22, 28, 85, 102, 109, 110, 111
Usnea barbata (or Bearded usnea), 10, 26, 27
Usnea florida, 48
Usnea hirta, 26
Usnea jubata, 10, 47
Usnea plicata, 27

Xanthomendoza (Gn.), 161
Xanthoparmelia (Gn.), 88
Xanthoria elegans (or Elegant sunburst lichen), 20, 132–3
Xanthoria parietina (or Yellow scale lichen/Parmélie des murailles), 18, 27, 49, 61, 102, 104, 109, 111, 173, 177, 200, 201, 205, Ill. 6